REGULAR FABRICS IN DEEP SUB-MICRON INTEGRATED-CIRCUIT DESIGN

REGULAR FABRICS IN DEEP SUB-MICRON INTEGRATED-CIRCUIT DESIGN

by

FAN MO and **ROBERT K. BRAYTON**

University of California, Berkeley

Library of Congress Cataloging-in-Publication

Title: Regular Fabrics in Deep Sub-Micron Integrated-Circuit Design
Author (s): Fan Mo and Robert K. Brayton
ISBN 978-1-4757-7934-9 ISBN 978-1-4020-8041-8 (eBook)
DOI 10.1007/978-1-4020-8041-8

Originally published by Kluwer Academic Publishers in 2004
Softcover reprint of the hardcover 1st edition 2004

Printed on acid-free paper.

TABLE OF CONTENTS

PREFACE

Regularity is a feature, which can provide better guarantees that the layout designed by a CAD tool is replicated in the fabrication. As the limits of the mask-making system (finite aperture of projection) are reached with smaller geometries, the actual layout patterns on the wafer differ from that produced by a CAD tool. Although pre-distortion can be added to offset some of the real distortions, the number of layout patterns generated by a conventional design flow can make this task take an unreasonable amount of time and generate an enormous data set. Beyond what optimal pre-distortion could do, regular circuit and interconnection structures can reduce variations further. Another motivation for using regularity is the timing closure problem, which arises because the design flow is sequential; early steps need to predict what the later steps will do. Inaccurate prediction leads to wrong decisions, which can only be discovered later, making design iteration necessary. Preventing such iterations is difficult, but use of regular structures, can make estimation much more accurate.

The following developments have been made to implement a timing-driven Module-Based chip design flow: (1) new regular circuit structures and their design methodologies, (2) a new integrated placement and routing algorithm that simplifies the standard-cell physical design, (3) a new block-level placement and routing algorithm with buffer insertion and (4) a multi-version approach allowing each module to carry several versions that can be selected by the block-level physical design algorithm. Test results show that this design flow can reach timing closure much faster than the Physical Synthesis flow (represented by several commercial synthesis tools) and achieve shorter clock cycle times.

ACKNOWLEDGEMENTS

We are grateful to members of the Nexsis Group at Berkeley, Philip Chong, Satrajit Chatterjee, Aaron Paul Hurst and Yinghua Li for many enlightening discussions. Also we gratefully acknowledge the help from Professor Qingjian Yu and Dr. Abdallah Tabbara, who were earlier members of the Nexsis Group. Several discussions on various research problems with Dr. Andreas Kuehlmann of Cadence Berkeley Labs, Dr. David Kung of IBM T.J. Watson, Dr. Charles Alpert of IBM Austin, Dr. Timothy Kam of Intel, Dr. Xiaojian Yang of Synplicity and members of the Constructive Fabrics Group of GSRC have helped to broaden the scope of this work as well as to refine our focus. We would like to thank Dr. Jun Ye and Dr. Yen-Wen Lu of Brion Technology for providing OPC pictures.

The work in this book was supported by the GSRC (grant from MARCO / DAPPA 98DT-660, MDA972-99-1-0001). We would also like to thank other partial sponsors of this research, including the California MICRO program and industrial supporters Fujitsu, Cadence, and Synplicity.

Chapter 1

INTRODUCTION

1.1. Deep Sub-Micron IC Design Flow and Regular Structures

The design of Integrated-Circuits (IC) is very complex; the complexity keeps growing as fabrication technology develops. Computer Aided Design (CAD) has become a key factor in successfully producing ICs with correct behavior and timing, and satisfactory yield. CAD is a set of computer programs, implementing algorithms that solve certain problems in IC design. Design flow is the organization of various algorithms and design styles to turn the IC design task into an automatic procedure. IC design has begun to benefit from the Deep Sub-Micron (DSM) technology, which offers higher density of transistor integration and higher performance. However, DSM technologies introduce many new problems [BRYA01, MO02], challenging the feasibility and efficiency of current CAD tools. One of the most critical issues is timing closure of the design flow, where the iteration between physical and logical CAD tools fails to converge to the desired timing behavior. The timing closure problem is due mostly to the growing importance of wiring effects. The separation of logic synthesis and physical design, two sequential stages in a traditional design flow, has become unacceptable since the two stages are tightly coupled by wiring effects. Timing convergence can no longer be easily guaranteed even if iterations between the stages are done. Various improvements have been made, among which the Physical Synthesis technique is popular [SYNO, CADE]. However the timing closure problem is far from being solved.

One approach to timing closure is the use of regular structures. In general, a regular structure involves the repetition of a basic pattern. Examples are memory arrays, dataflow modules, PLAs, and gate arrays. Regular structures have several advantages. First, they are more manufacturable since their basic pattern can be more precisely corrected optically (OPC); more computation can be spent on OPC while neighboring cells can be better predicted to further enhance the OPC. Another advantage of regular structures is that they can have more predictable timing properties. This is due partly to their more predictable manufacturing, but perhaps more due to their regular and predictable wiring. Regular routing means that the routes that the wires take are typically known in advance and thus their delays are much more predictable earlier in the design cycle. Depending on the regular structure used, routing can be known even at the logic synthesis stage or at least at the placement stage. This means that wire delays can be predicted better at these early stages, alleviating the timing closure problem. Regular routing also results in a reduced number of vias, which are a major source of unpredictability in DSM because of large variations in via resistance during the manufacturing process.

This book explores various regular structures, both logical structures like PLAs as well as local and global routing structures. The purpose is to offer alternate design flows and to provide reasonable estimates for assessing the costs and trade-offs of using such structures. We provide several new regular logical structures along with associated logical and physical synthesis algorithms. Finally, to provide a global assessment, a new "module-based" design flow is presented, which assembles regular modules and places and routes them at the chip level using a new global routing scheme.

The Module-Based design flow takes a very different approach to reaching timing closure. Although it is not claimed that the timing closure problem is completely solved by this flow, it does produce better results than the Physical Synthesis flow, both in terms of the minimum achievable clock cycle as well as a much shorter run time than the Physical Synthesis flow. This allows users to obtain feedback more quickly and make necessary design changes without waiting a long time to get a failure notice.

The Module-Based design flow consists of the following developments:

(1) new circuit structures and their synthesis methodologies,
(2) a new placement and routing algorithm that simplifies standard-cell physical design,
(3) a new block-level placement and routing algorithm with buffer insertion, and
(4) a multi-version approach allowing each module to carry several versions that can be selected during the block-level physical design algorithm.

Test results show that this design flow can reach timing closure much faster than the Physical Synthesis flow, plus it can achieve shorter clock cycles.

1.2. Outline

Chapter 2 gives the preliminaries, including an overview of CMOS technology, delay modeling, the timing closure problem and other DSM issues. Since a design flow is always associated with one or more circuit structures, in Chapter 3 we discuss various regular structures and their design methodologies including logic and physical synthesis methods for each. These structures include the widely used (but less regular) standard-cell and Programmable Logic Arrays (PLAs), in addition to several new circuit structures. In Chapter 4, a block-level placement and regular routing scheme called Fishbone is presented. The Fishbone wiring structure is a simple regular scheme which has the benefit of easy and efficient routing which can be done during placement. Hence wiring delays can be precisely estimated, leading to a placement that meets timing, while having the routing completed at the same time as the placement. An extension to Fishbone that enables simultaneous buffer insertion is also discussed. This is necessary for establishing the correct timing on the global wires. The entire Module-Based design flow is presented in Chapter 5, and comparisons are made with a typical Physical Synthesis flow. Chapter 6 concludes.

Chapter 2

PRELIMINARIES

2.1. Overview

In this chapter, a quick overview of CMOS technology and basic design automation techniques are given. Several circuit configurations are described: static CMOS, static NMOS and dynamic CMOS. These configurations are widely used in current IC designs and are also fundamental elements of the new structures we propose in the next chapter. Delay formulations for gate and wire are also given. These will be adopted throughout the discussion in the following chapters. Finally DSM challenges are addressed.

2.2. CMOS Technology

2.2.1. Transistors

The basic component of Integrated Circuits (IC) is transistor [WEST94, SZE85, ONG84, YANG88, MULL86, STRE80]. Interconnected transistors form logic functions to meet the system requirements. A chip, or an IC unit, is a multiple layer structure. The bottom layers, usually silicon with different types of doping, build the electronic devices like transistors; the upper layers, usually metal, form the interconnections between the electronic devices. During fabrication, the layers are grown from the bottom [CLEI99]. Each layer is associated with a mask, on which patterns indicate which parts of the layer should be kept (or removed). The patterns on the mask are duplicated physically on the wafer through optical and chemical processes [PLUM00]. The task of Computer Aided Design (CAD) for IC is to translate system specification to masks.

CMOS stands for "Complementary Metal Oxide Silicon". It is today's most widely used IC technology. The basic element of CMOS technology is the MOS transistor or MOSFET (Metal-Oxide-Silicon Field Effect

Transistor). Figure 2.1 illustrates the physical structure and the schematic view of a MOS transistor.

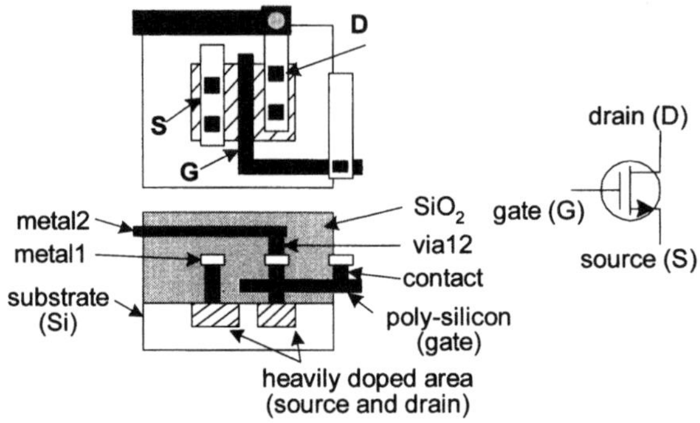

Figure 2.1: A MOS transistor

A transistor has three terminals, drain (D), source (S) and gate (G). The body of the transistor is built on the silicon substrate. A layer of poly-silicon (poly), being stacked on a very thin layer of silicon-dioxide, forms the gate terminal. The source and drain are the regions on the substrate that are implanted or doped with impurities. To connect a transistor to others, which are built somewhere else on the substrate, multiple metal layers (here only two are shown; in reality, there might be more than six for DSM) are used. The metal layers are connected by vias, which are made of special metals. Metal layer 1 is connected to the substrate or the poly-silicon by contacts, which are also made of special metals. The space between objects is filled with silicon-dioxide, a kind of isolating material. To fabricate the transistor shown in the figure, the CAD tool should produce at least the following masks: substrate doping or implantation, poly-silicon, contact, metal1, via12 (via connecting metal 1 and 2) and metal2.

The transistor can be viewed as a switch with two mutually exclusive states, on and off. Gate is the controlling terminal. When high voltage is applied to the gate, there exists a conductive path, sometimes called “channel”, between source and drain. When the gate is low, the conductive path between source and drain is cut off. The transistor shown in Figure 2.1 is an NMOS (N-type MOS) transistor. There is a complementary type called a

PMOS (P-type MOS) transistor. The PMOS transistor works as a switch too, but a low controlling signal causes the switch to conduct.

2.2.2. Gates

To build logic functions, transistors are connected to form gates. Here "gate" refers to a circuit unit that implements certain logic functions (it happens to have the same name as the "gate" terminal of the MOS transistor). Three typical gate configurations are discussed: static CMOS, static NMOS and dynamic gates.

2.2.2.1. Static CMOS gates

The serializing and paralleling of NMOS and PMOS transistors create CMOS gates [KANG95]. The NMOS transistors build a conductive path to the ground, and the PMOS transistors build a conductive path to power. The conductivities of the transistors are controlled by input signals. The two paths join at the output pin of the gate. At any moment, one and only one path is conductive, which ensures a stable output. To make this possible, the controlling of the two paths should be logically complemented (mutually exclusive and complete).

As shown in Figure 2.2(a), a CMOS inverter is composed of a PMOS and an NMOS transistor. When input A is high, the NMOS transistor is on and the PMOS is off, so the output Z is pulled down by the NMOS to ground (or zero Volts) via its conductive path. When the input A is low, the NMOS turns off and the PMOS turns on, so Z goes high. Figure 2.2(b) outlines the timing model of the inverter. An inverter has an input capacitance C_I and an output capacitance C_O. The two resistors, R_P and R_N, model the PMOS and the NMOS switches, respectively. The external load capacitance C_L is the sum of the input capacitances of all the gates the inverter drives. To be more precise, C_L should include the wire capacitances, because the inverter needs some metal wires to connect to the gates it drives. An "on" transistor can be approximated as a resistor. Charging C_L and C_O through resistor R_P and discharging them through R_N both take time, which generate transistor delays. The transistor delays depend on many factors, among which the transistor size and the load are the most important. The larger a transistor, the smaller the resistor and the faster it switches. The larger the load is, the slower the switching. If R_P and R_N are equal, the two resistors and the switch can be simplified to a voltage source that drives one resistor R_O=R_N=R_P. A typical waveform of the CMOS inverter is illustrated in Figure 2.2(c). The delays are shown as the intervals between the transition edges of the input and output. In general delay might be different for low-to-high and high-to-low transitions. However, the size ratio of the PMOS and NMOS transistors can be designed such that *delay*=t_{HL}=t_{LH}.

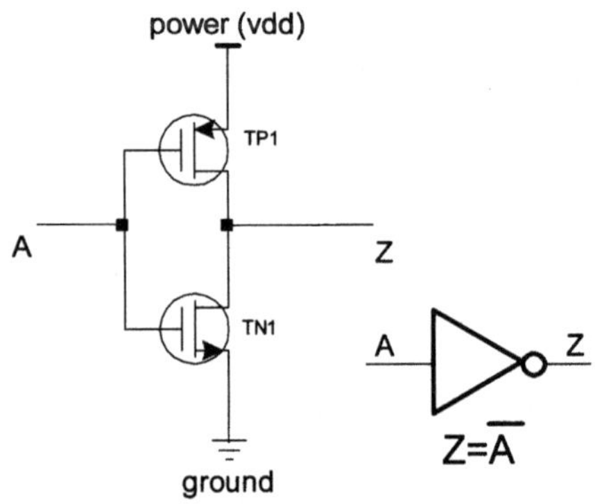

(a) CMOS inverter gate

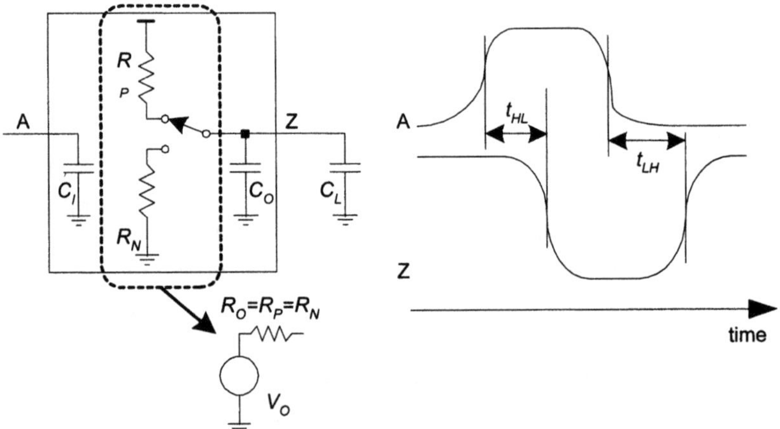

(b) timing model (c) transient behavior

Figure 2.2: CMOS inverter

To obtain more complex logic functions, transistors can be connected in parallel or in serial. Fundamental logic operations, such as INV (¯ , inversion), AND (&), OR (+), XOR (^, exclusive-OR) etc., have their transistor-level representations. A logic function might be prefixed with an "N" to indicate that its output is inverted, e.g. an NAND gate, as illustrated in Figure 2.3(a). In general, as shown in Figure 2.3(b), a CMOS gate consists of a network of NMOS transistors between the output and ground, which form the pull-down logic, and a network of PMOS transistors between the output and power to form the pull-up logic. The pull-up and pull down parts are logically complemented; hence conductive pull-down and pull-up paths do not coexist. In this sense, static CMOS gates do not consume *static* power.

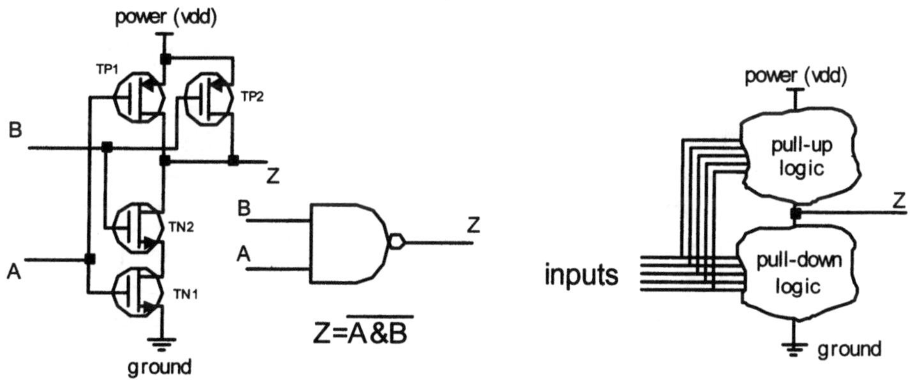

(a) 2-input NAND gate (b) general CMOS gate

Figure 2.3: CMOS gate

A special kind of gate is called a buffer. It is two inverters in a chain. A buffer expresses an "=" function, which may sound useless in building circuits, but it isolates heavy load from weak driver when inserted into a wire. Buffer insertion is a powerful means of handling timing problems and is discussed in Section 2.2.4.2.

2.2.2.2. Static NMOS gates

Before CMOS technology became popular, only NMOS transistors were used to build logic gates [HODG03].

Figure 2.4(a) illustrates a typical NMOS inverter, which is simply a NMOS transistor (TN1) connected to a depletion-type NMOS transistor. The depletion-type NMOS transistor has a zero or negative threshold voltage such that even if its gate and source are zero biased, the transistor is on. This makes TN0 form the pull-up function, as if it is a resistor. When A is low, TN1 is off and Z is pulled to high through TN0. When A is high, TN1 is on and it overcomes TN0 such that Z becomes low. Figure 2.4(b) shows how general logic functions are constructed with the NMOS configuration. The lower part of the NMOS inverter is replaced by a network of NMOS transistors, the input signals determining the existence of a conductive path from Z to ground.

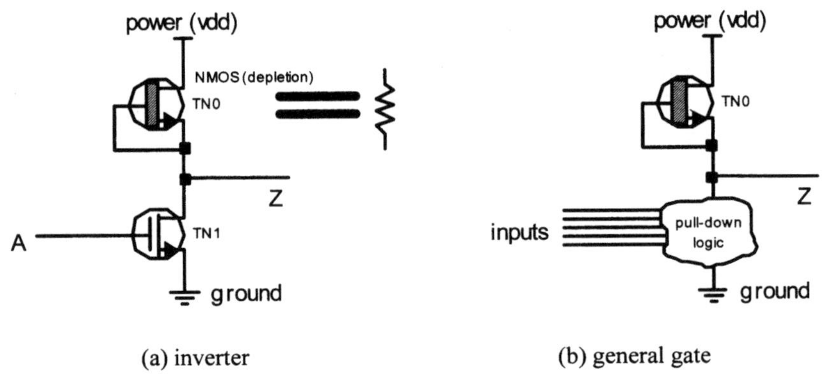

Figure 2.4: NMOS gate

NMOS gates are compact compared with CMOS gates, because a NMOS gate saves all the PMOS transistors. In addition, PMOS transistors are often laid out such that they are larger than NMOS transistors to balance the rising and falling delays [HODG03]. Therefore the area saving of NMOS gates is significant. However static NMOS gates suffer in delay. The choice of the sizes of the NMOS transistors and the sizes of the pull-up transistor is not easy. Consider the inverter in Figure 2.4(a). As the function outputs low, the pull-up transistor TN0 battles with TN1. To ensure reaching a low output level quickly enough, TN1 should strongly overcome TN0. In contrast, when TN1 turns off and the level of Z rises, it is the driving capacity of TN0 that determines the rise time. The two factors are obviously contradictory. Another disadvantage of static NMOS gates is the power dissipation. A static NMOS gate consumes static power when its output is low.

To mitigate the delay and power problems while preserving the compactness of the NMOS gates, a dynamic configuration is introduced.

2.2.2.3. Dynamic configuration

The CMOS and NMOS gates described above both belong to the static configuration classification, in the sense that the output responds immediately (after some gate delay) to the change at the inputs. The dynamic configuration only requires that the output respond within a certain time period [HODG03]. This is somewhat similar to static NMOS, except that the pull-up transistor is replaced by a pair of pre-charging-control transistors.

(a) inverter (b) general gate

Figure 2.5: Dynamic gate

A dynamic inverter is illustrated in Figure 2.5(a), in which TP0 and TN0 control the state of the gate, either "pre-charge" or "evaluate". When $\overline{\text{precharge}}$/evaluate is low, TN1 is isolated from ground because TN0 is off, and the output Z is pulled up to high by TP0. When $\overline{\text{precharge}}$/evaluate becomes high, TP0 is off and TN0 is on. TN1 starts to evaluate the logic function controlled by input A, and Z responds accordingly. In a general dynamic gate [DHON92], as illustrated in Figure 2.5(b), the evaluation is done through a network of NMOS transistors.

There are many variations of the dynamic configuration [HODG03], but all require a $\overline{\text{precharge}}$/evaluate signal that controls the operation of the gate. If dynamic gates are cascaded, timing validity needs to be considered for the generation of the $\overline{\text{precharge}}$/evaluate signals for each gate in the chain. In other words, a dynamic gate cannot start to evaluate until all its input signals are evaluated. Several Programmable Logic Array (PLA) based circuit structures described in the next chapter make use of the dynamic configuration.

2.2.3. CMOS storage components

Gates are reactive to their inputs. Other components have a sort of memory function, reacting to the inputs under certain conditions. Such storage components can store previously inputted information. A typical storage component is register, as shown in Figure 2.6. The pins of a register include clock C, data-in D and output Q. The function of a register can be summarized as "whenever C rises, Q copies D and hold it until the next rising edge of C". This kind of edge-triggered register is called a "flip-flop".

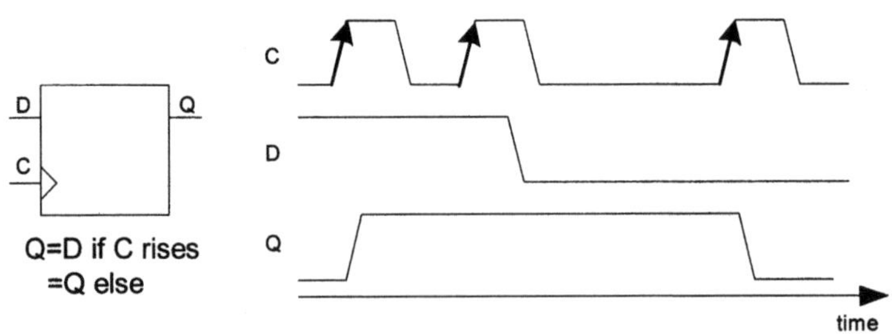

Figure 2.6: Register and its operation

Memory components such as Random Access Memory (RAM) also provide storage, which are arrays of storage units. Accessing a bit of the memory (read or write) is at least several times slower than accessing a register. On the other hand, memory components are more compact in layout.

2.2.4. Interconnections

2.2.4.1. Wire delay

The Elmore delay formulation is the wire delay model we adopt [ELMO48]. The Elmore model formulates the propagation delays as a tree with output pin being the root. At each branching point including the root, the delay is the resistance of the last wire segment times the total accumulated capacitance of the sub-tree rooted at the branching point.

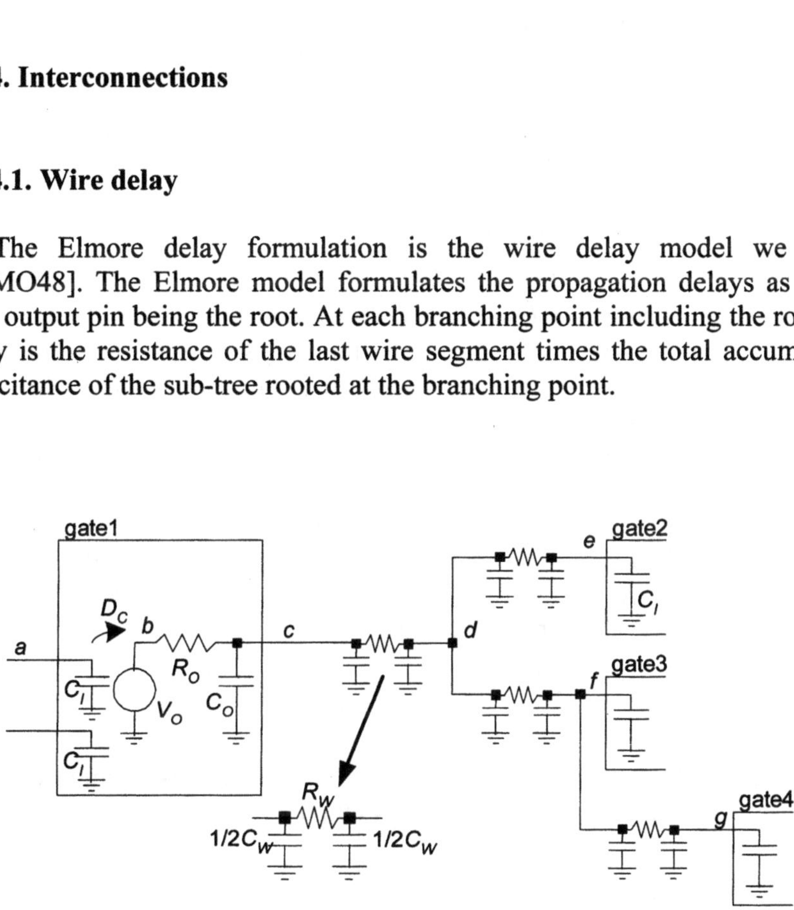

Figure 2.7: Gate delay and wire delay models

In Figure 2.7, the driving gate is gate1, the loading gates are gate2 to 4. Gate2, gate3 and gate4 contribute their input capacitances to the load of gate1. For a wire segment with length L, a π-shaped RC structure is assumed, as shown in the figure. The capacitance and resistance of the segment are:

$$C_W = L \cdot C_{//} \text{ and } R_W = L \cdot R_{//},$$

where $C_{//}$ and $R_{//}$ are the capacitance and resistance of the unit wire length respectively. We examine the propagation delay from point *a* to point *f*. The delay is the sum of the delays of the segments:

$$D(a \to f) = D(a,b) + D(b,c) + D(c,d) + D(d,f).$$

$D(a,b)=D_C$ is the intrinsic gate delay, which is load independent. The delay from *b* to *c*, also part of the gate delay but load dependent, is:

$$D(b,c) = R_O[C_O + C_{ACC}(c)],$$

in which, $C_{ACC}(c)$ is the accumulated capacitance at node *c*, to be clarified shortly. $D(c,d)$ and $D(d,f)$ are the delays on wire segment *c-d* and *d-f* respectively:

$$\begin{aligned} D(c,d) &= R_W(c,d)C_{ACC}(d) \\ D(d,f) &= R_W(d,f)C_{ACC}(f) \end{aligned},$$

in which, $C_{ACC}(d)$ and $C_{ACC}(f)$ are the accumulated capacitances at node *d* and *f*, and $R_W(\cdot,\cdot)$ is the π-shaped resistance of the wire segment. The formulation of $C_{ACC}(\cdot)$ is essential in the Elmore model. A net is a tree $\{P,E\}$, where *P* is the node set and *E* is the edge set. The output pin is the root, and other nodes are either the input pins of the loading gates or the branching points. Each edge is denoted by a node-pair (i,j) and the signal direction is from *i* to *j*. An edge corresponds to a wire segment, so it is associated with a π-RC, denoted by $C_W(i,j)$ and $R_W(i,j)$ and both are linear with the length of the segment. Define for each node a variable $C_{ACC}(i)$ called the accumulated capacitance. The value of $C_{ACC}(i)$ is recursively defined:

$$C_{ACC}(i) = \sum_{\{h,i\}\in E}\left[\frac{1}{2}C_W(h,i)\right] + \sum_{\{i,j\}\in E}\left[\frac{1}{2}C_W(i,j) + C_{ACC}(j)\right] + C_I^?$$

for non-leaf nodes. The term $C_I^?$ equals to C_I if this node is an input pin, or 0 otherwise. For leaf nodes which are always input pins,

$$C_{ACC}(i) = \left.\frac{1}{2}C_W(h,i)\right|_{\{h,i\}\in E} + C_I.$$

This model can be summarized as "the accumulated capacitance is the sum of all capacitances of the downstream nodes". So at the output pin (node *c* in the example) or the root of the tree, the accumulated capacitance is the sum of all π-capacitances and all input pin capacitances. Hence,

$$C_{ACC}(e) = \frac{1}{2}C_W(d,e) + C_I$$

$$C_{ACC}(g) = \frac{1}{2}C_W(f,g) + C_I$$

$$\begin{aligned} C_{ACC}(f) &= \frac{1}{2}C_W(d,f) + \frac{1}{2}C_W(f,g) + C_{ACC}(g) + C_I \\ &= \frac{1}{2}C_W(d,f) + C_W(f,g) + 2C_I \end{aligned}$$

$$\begin{aligned} C_{ACC}(d) &= \frac{1}{2}C_W(c,d) + \frac{1}{2}C_W(d,f) + C_{ACC}(f) + \frac{1}{2}C_W(d,e) + C_{ACC}(e) \\ &= \frac{1}{2}C_W(c,d) + C_W(d,f) + C_W(f,g) + C_W(d,e) + 3C_I \end{aligned}$$

$$\begin{aligned} C_{ACC}(c) &= \frac{1}{2}C_W(c,d) + C_{ACC}(d) \\ &= C_W(c,d) + C_W(d,f) + C_W(f,g) + C_W(d,e) + 3C_I \end{aligned}$$

Then the delay on each segment can be derived and the total delay is:

$$\begin{aligned} & D(a \to f) \\ &= D(a,b) + D(b,c) + D(c,d) + D(d,f) \\ &= D_C + R_O\left[C_O + C_W(c,d) + C_W(d,f) + C_W(f,g) + C_W(d,e) + 3C_I\right] \\ &\quad + R_W(c,d)\left[\frac{1}{2}C_W(c,d) + C_W(d,f) + C_W(f,g) + C_W(d,e) + 3C_I\right] \\ &\quad + R_W(d,f)\left[\frac{1}{2}C_W(d,f) + C_W(f,g) + 2C_I\right] \end{aligned}$$

Except for a constant part D_C, the total delay depends on the wire RC parameters and the input pin capacitances of the loading gates. The input pin capacitances are known, but the wire RC's are determined by the actual net topology, wire widths and routing layers. An interesting phenomenon is that the delay along the wire segments *c-d-f* also depends on the segments *d-e* and *f-g*. Since the π-shaped R and C are both linear functions of the length of the wire segment, the wire related delay grows quadratically as the wire length increases. In the above derivation, we assume all the loading gates have the same input capacitance C_I. In real circuits, C_I is proportional to the gate size. As mentioned earlier, to make a gate switch faster, or a smaller D_C, the gate size should be large. But C_I becomes large as a result. So to make a gate switch faster by increasing its size, we actually increase the load to the previous gates. Therefore the optimization of the delay of a chain of gates is an interesting and important problem in CAD [HU03].

2.2.4.2. Buffer insertion

Buffer is a kind of gate that fulfills the "equal" function. Inserting buffer(s) in a wire does not change the logic function, but provides a powerful means of reducing wire delay [CONG01a, CONG01b]. The main idea is to break a long wire into short segments where buffers are inserted at the break points. The short wire segments reduce the quadratic wire delay effect; although the inserted buffers introduce extra delays because of their intrinsic and load dependent delays. Hence the problems are how many buffers are inserted and where. Here we give a simple description of the buffer insertion for a two pin net (one driver and one load, or fanout=1). In Chapter 4, the buffer insertion problem in block-level physical design will be detailed.

Suppose the inserted buffers have the same capacitance C_I as the input pins and the same driving resistance R_O as the output pin, and they divide the wire into N segments with equal length. By using $C_{//}$ and $R_{//}$, the unit length wire capacitance and resistance, the un-buffered wire delay can be expressed as:

$$D = R_O(LC_{//} + C_I) + LR_{//} \cdot \left(\frac{LC_{//}}{2} + C_I\right),$$

in which, L is the total wire length. The delay after $(N-1)$ buffers are inserted becomes:

$$D' = N \cdot \left[R_O\left(\frac{LC_{//}}{N} + C_I\right) + \frac{LR_{//}}{N}\left(\frac{LC_{//}}{2N} + C_I\right)\right] + (N-1)D_{BUF}$$

where D_{BUF} is the intrinsic buffer delay. Comparing D and D', we see that the total delay is decreased by

$$\Delta D = D - D' = (N-1)\left[\frac{L^2 R_{//} C_{//}}{N} - (R_O C_I + D_{BUF})\right].$$

A positive value is necessary for meaningful buffer insertion. Thus delay cannot be arbitrarily improved by inserting more and stronger buffers. Figure 2.8 shows the wire delay as a function of the number of wire segments (the number of buffers plus one). The parameters used in making the plot are: $C_{//}$=1.1×10^{-4}pF/μm^2, $R_{//}$=5.0×10^{-2}Ω/μm, C_I=0.001pF and R_O=100.0Ω. The length of the wire is 5.0mm, and the buffers are evenly inserted along the wire.

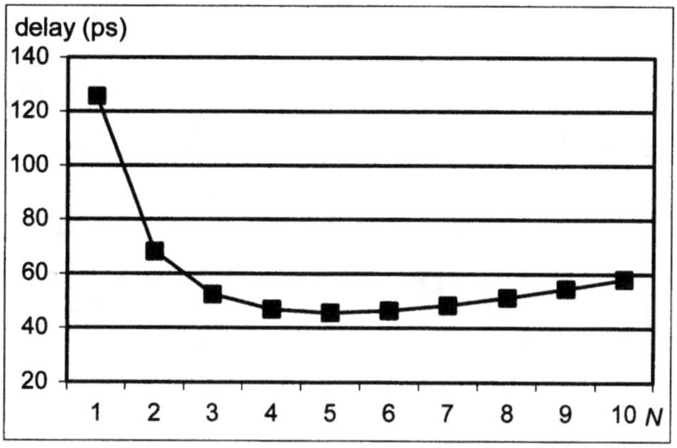

Figure 2.8: Delay versus number of wire segments (N) evenly split by buffers

The analysis is more complicated for multiple fanout nets. In addition, buffer insertion may increase the chip area, which may in turn change the wire lengths and thus the wire delays, making the prediction of the wire length even harder. Also, to insert a buffer, vias need to be introduced to bring the net, most probably running on higher metal layers, down to the substrate for buffering and bring the buffered signal back to higher layers. The number of vias used to connect to the buffer, and thus the resistance, is hard to predict. Furthermore, the buffer locations may not be arbitrarily chosen, because existing gates and modules may have already occupied a location.

2.2.4.3. Net topology

A net is a set of pins that are logically connected. A net must have at least two pins, and one of the pins must be output pin (driver) and the rests are input pins. A bi-directional pin can be either output or input, controlled by some other signal. A net can have more than one bi-directional pin. But to prevent conflicts, one and only one pin of a net drives the rest pins at any time. We do not consider the case that bi-directional pins appear in the nets.

A general net may contain more than one input pin (load), such that intermediate points may exist on the net generating branches of the wires.

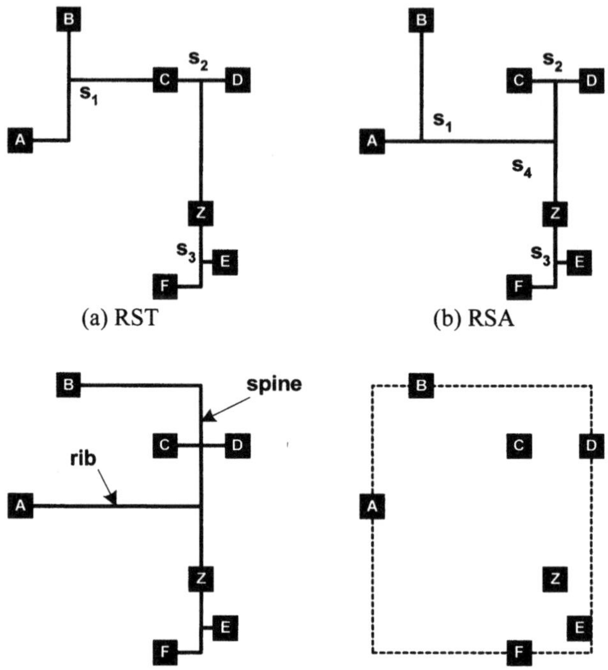

Figure 2.9: Net topologies

Various net topologies exist [SHER93, SHER95]. Some popular and useful net topologies are illustrated in Figure 2.9. The net in the figure is composed of output pin Z and input pins A to F. Criteria are needed to evaluate the quality of the net topology. The most widely used criterion is wire length. Short wire length is likely to result in easy routing, because of the low utilization of the routing resource. For two pin nets, reducing wire length means shorter wire delay. A Rectilinear Steiner Tree (RST) [SHER93], as shown in Figure 2.9(a), is a net topology that provides the minimum wire length for a set of pins of a net. The non-pin nodes in the tree topology, s_1, s_2 and s_3 in the figure, are called Steiner points. However RSTs do not always provide minimum length between any output-input pin pair, for instance the Z-A connection in Figure 2.9(a). Rectilinear Steiner Arborescence (RSA) [CONG97], as illustrated in Figure 2.9(b), is an improved Steiner Tree that each output-input pin connection has minimum length. Unfortunately both of these two topologies are computationally hard. Figure 2.9(c) illustrates a so-called Spine topology. The output pin is on a wire segment called the "spine" and all input pins connect to the spine via "ribs", segments orthogonal to the spine. The construction of the spine for a given set of pins is trivial. Like RSAs, the output-input pin lengths are always the minimum. To quickly estimate the wire length of a net without topologically constructing the net, a Half Perimeter (HP) model is usually adopted. As illustrated in Figure 2.9(d), the HP for a net is the half perimeter of the minimum bounding box surrounding all pins. Strictly speaking, HP is not a net topology, but since a net is physically constructed with RST, RSA, Spine or other topologies, the delay analysis described above can be performed.

Following are some important points regarding the net topologies:

- For a given set of pins and their locations, the RST and RSA may not be unique.
- In reality, the cells carrying the pins are movable; so minimizing wire length or delay is not only the job of choosing a net topology but also placing the cells at good positions.
- The topologies only tell the X/Y information. The Z information, the metal layer a wire segment is placed on, is not known. Whenever a wire switches layers, a via (or vias if it switches to a non-neighboring layer) is introduced. This is usually determined by the router. However delay computations may need layer information to get precise results, because different layers and vias have different physical parameters. Unfortunately, this leads to the conclusion that only after routing is precise delay computation possible.
- Buffer insertion may further change the quality of a net topology.

2.3. IC Design and System-On-a-Chip

2.3.1. System and constraints

A digital circuit system can be regarded as a state machine. The storage of the states is implemented with storage components. Digital systems are classified into two categories, synchronous and asynchronous. Asynchronous systems are not in the scope of our discussion.

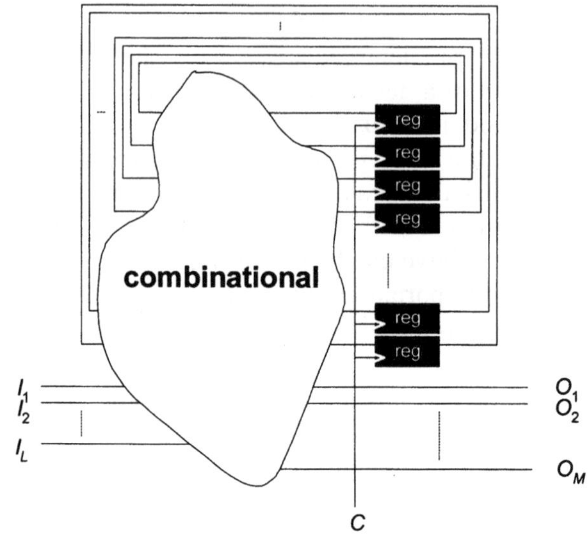

Figure 2.10: A state machine

A synchronous system can be viewed as a state machine, as shown in Figure 2.10. The state machine contains two parts: a combinational part and a sequential part composed of Z registers clocked by signal C. The state is described by a binary value stored in the registers that contains Z bits. Obviously the system has up to 2^Z different states. The state is migrated to the new state at the clock edge. Between two clock edges, the combinational part, composed of gates, computes the new value of the state, and the new state is ready at the inputs of the registers when the next clock edge arrives. The external inputs to the system are labeled $I_1, I_2, \ldots, I_L$, and the outputs are O_1, $O_2, \ldots, O_M$. Denote the system clock cycle by K. The behavior of the synchronous system can be formulated as:

$$S(t+K)=f(S(t), I(t)) \qquad 2^Z \times 2^L \to 2^Z$$
$$O(t)=g(S(t), I(t)) \qquad 2^Z \times 2^L \to 2^M$$

in which, $S()$ is the state, $f()$ is the state transition function, and $g()$ is the output function. In reality, a large system is normally described as a network or hierarchy of many small state machines, because first, hierarchical description is more natural; second, small state machines may be more efficiently implemented with CAD tools. The interconnection of small state machines can result in a single state machine with a huge number of states, although they are functionally equivalent.

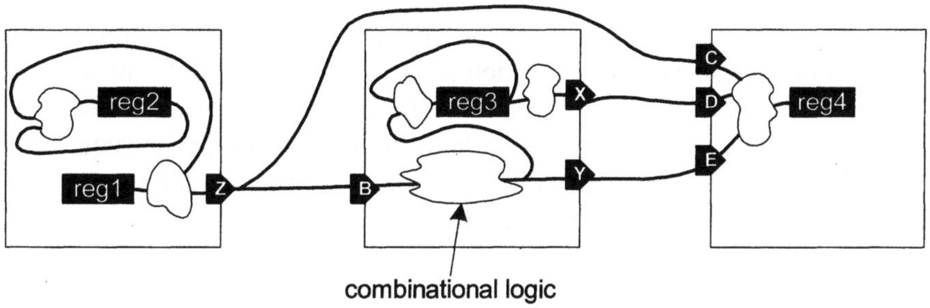

Figure 2.11: Combinational paths

To measure the performance of a synchronous system, we need to consider combinational paths, defined as a path starting from a register and ending at a register. In Figure 2.11, various combinational paths are shown. A combinational path can start from a register and end in a register in the same module, e.g. paths feeding register 2 in the left module. A combinational path may cross modules, like the one starting from register 1, passing through pins Z, B, Y and E, and ending at register 4. Note that a path can have branches along the way, and multiple paths may share some common segments. In addition, more than one path can exist between a pair of registers, e.g. register 1 and 4 in the figure.

Timing constraints are normally expressed in terms of "arrival times" and "required times". These are defined on the signal nodes of the circuit such as register pins, the pins of the modules and pins of the I/O ports. Assume all registers update their output values at time 0, i.e., the arrival times of the register output pins are 0. All register input pins should receive the new input no later than $t_{REQ}=n(path)\cdot K$, where K is the clock cycle and $n(path)$ is an integer representing the number of cycles allowed for the path reaching that register. If $n(path)=1$, the path is called a single-cycle path. Satisfaction of the timing constraints can be stated as:

$$t_{ARR}(node) \leq t_{REQ}(node)$$

for all nodes. For a path starting from or ending at the environment through an I/O port, we simply assume that there is a register in the environment.

Figure 2.12 gives an example of the usage of the arrival and required times. When considering a path starting from the environment, e.g. the path through IPAD1 in the figure, we can imagine that it starts from external register rege1. The arrival time at the input I/O port ($t_{ARR}=0.6K$), must be provided in the specification, taking into account the wiring delay in the environment. Similarly, for output port OPAD1, the required time t_{REQ}=0.9K must be given. Paths coming out of or entering hard macro-cells, like hard-macro1, need to get arrival times on the output pins and set required times on the input pins. These numbers are determined by the internal path delays of the module and given in the specification. In the rest of the circuit, the dotted region in the figure, the register input and output pins are given required and arrival times, respectively. In Chapter 5, the required/arrival times will be extensively used in a timing driven design flow.

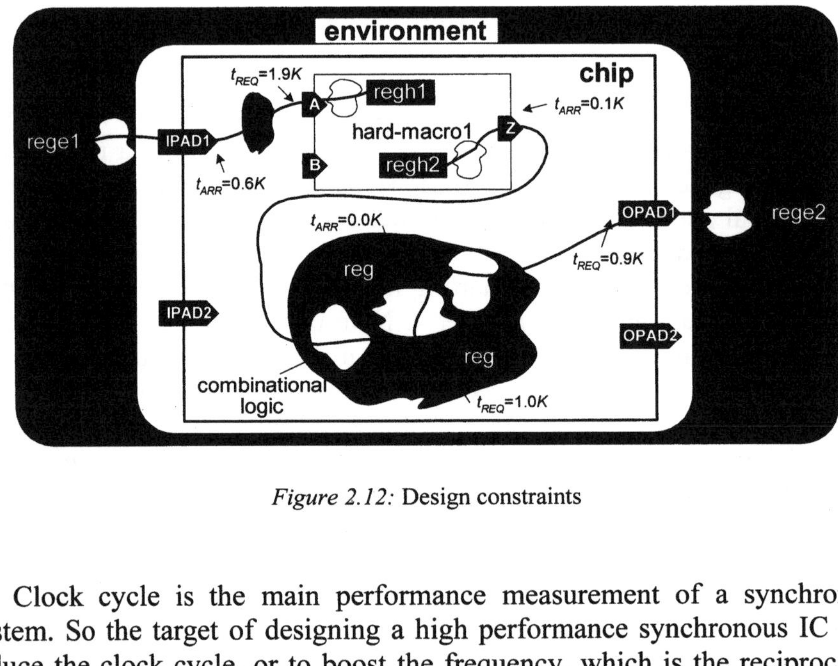

Figure 2.12: Design constraints

Clock cycle is the main performance measurement of a synchronous system. So the target of designing a high performance synchronous IC is to reduce the clock cycle, or to boost the frequency, which is the reciprocal of the clock cycle.

Finally we point out that path delay should include the wire delay. In modern IC design, wire delays may even dominate cell delays.

2.3.2. Design styles

Design style is the way a design is implemented, associated with certain circuit structures, among which standard-cell and macro-cell are popular. In Figure 2.13, three major styles in today's IC design are illustrated.

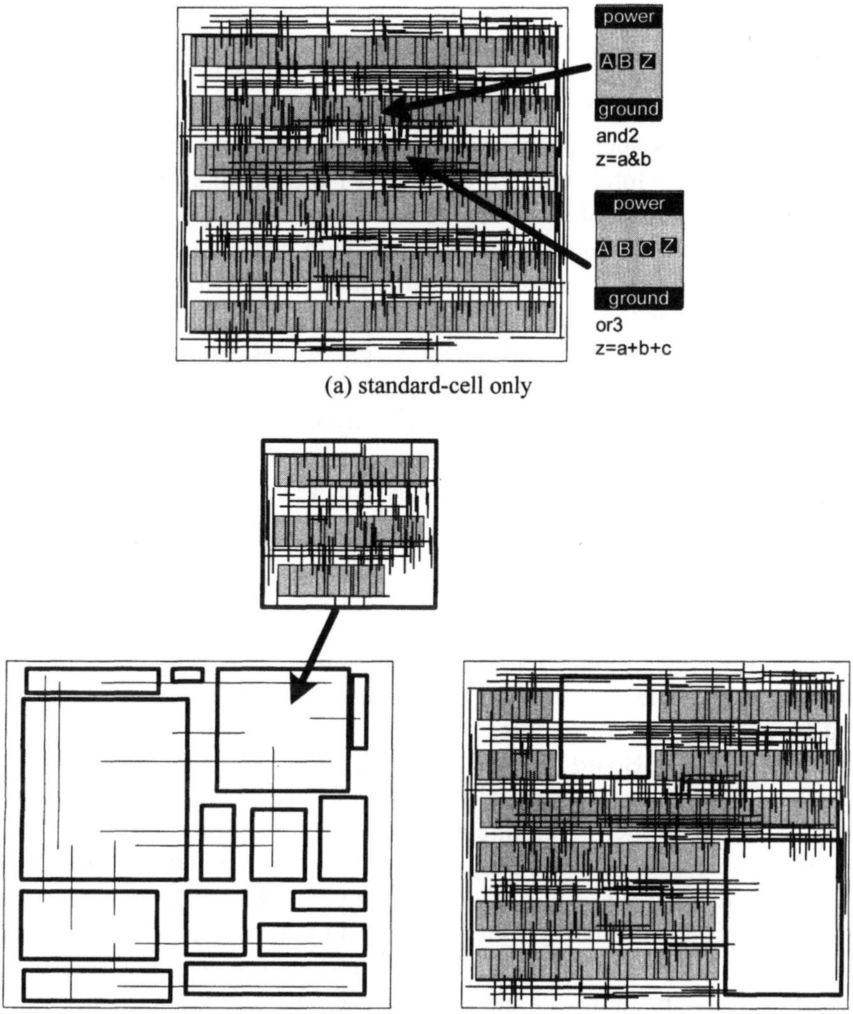

(a) standard-cell only

(b) macro-cell only

(c) mixed standard/macro-cell

Figure 2.13: Three design styles

In Figure 2.13(a), a pure standard-cell design is shown. Standard-cells, to be detailed in Chapter 3, are small pre-designed circuit modules that implement simple logic functions. All standard-cells in a design have the same height, so that they can be easily placed in rows. When the standard-cell design style is adopted, the design task is to map the input netlist to a netlist of standard-cells provided in a library, and then place the cells in the layout and construct their interconnections. The fine granularity of the standard-cell provides flexibility in the mapping and the layout design phases. In Figure 2.13(b), a macro-cell design is shown. Unlike standard-cells, macro-cells are modules with coarse granularity; they are usually several orders of magnitude larger than standard-cells, implementing more complex logic functions. Typical macro-cells include memory modules that are generated by some memory compilers, Intellectual Property (IP) blocks like embedded micro-processors and analog blocks like Phase Lock Loop (PLL). Of course a macro-cell can be a module that is designed with standard-cells. Pure macro-cell design involves much less components in the layout than the standard-cell only style. However this does not necessarily lead to simpler layout design, due to the heterogeneous shapes and sizes of the macro-cells. A mixed standard/macro-cell design is shown in Figure 2.13(c), which consists of both macro-cells and standard-cells in a single layout. Standard-cells, which are much smaller than macro-cells, are placed in the space left by macro-cells. The previous two styles can be regarded as special cases of this style.

2.3.3. System-On-a-Chip

When the integration density of IC's was low, an IC could only implement part of the system function. To build complex systems, many chips might be needed; and these chips were integrated on a board, leading to the term "System-On-a-Board (SOB)". Multi-Chip-Module (MCM) is another type of integration, putting several raw chips in one package. But this is only a "tiny" SOB. With increasing integration density due to the shrinking feature sizes, the whole system can be implemented on a single chip. The so-called System-On-a-Chip (SOC) is not a "very-tiny" SOB, but a new concept in electronic design [BEDN02]. One of the major advantages is that now the interconnections are also integrated and thus much shorter and denser than those in SOB's. This allows higher speed to be achieved. Another advantage is the reduction of packaging and board cost, because everything is now on a single chip. Unfortunately many new problems accompany the advantages. For example, the complexity of the chip grows, which must be dealt with by the design tools. Timing and power problems are both different from and more difficult than before.

It is interesting to see SOC designers are also pursuing higher and higher speeds, using the fact that an SOC can be faster than its SOB counterpart.

2.4. Design Automation

Electronic Design Automation (EDA) is a software approach that automatically, or with little human intervention, translates the initial logic network to a layout.

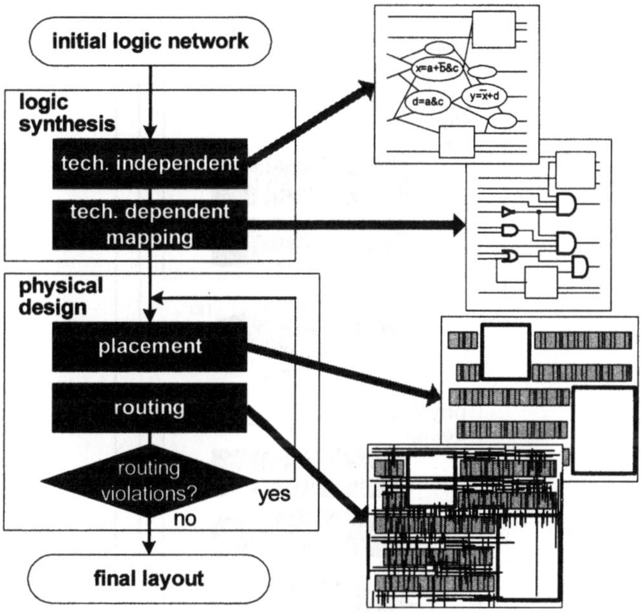

Figure 2.14: A simple design flow

Figure 2.14 shows a simple design flow, containing two stages: logic synthesis and physical design. The task of logic synthesis is to translate the initial logic network to a network that is composed of macro-cells and standard-cells. The logic synthesis can be further divided into a technology independent stage and a technology dependent stage [BRAY90]. The technology independent stage employs various Boolean transformations to reduce the size and/or delay of the circuit. The netlist produced by the technology independent optimization is given to the technology dependent stage, which has a more popular name "technology mapper". The technology mapper translates the Boolean network into a network of cells, while maintaining the same functionality. The cells are chosen from a library of pre-designed and pre-characterized modules called the standard-cells. The physical design stage works on the mapped netlist. It also contains two parts: placement and routing [SHER93]. Placement finds the positions for each component on the chip and routing determines the structures of the interconnections. The router may fail due to routing violations, in which case placement needs to be restarted with some parameters adjusted.

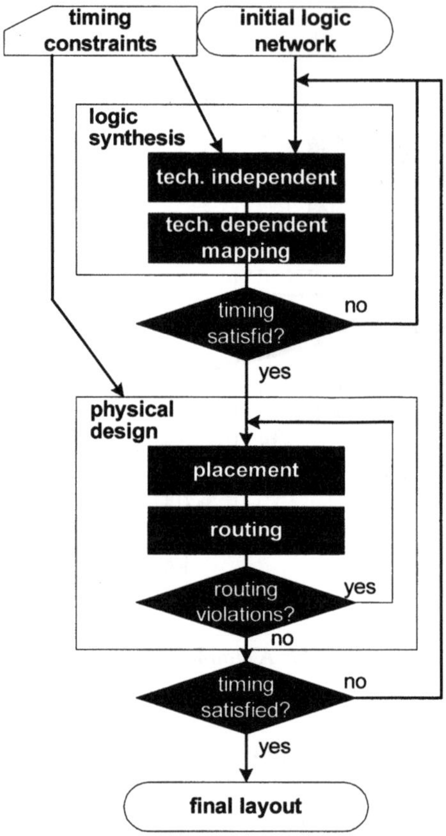

Figure 2.15: A timing-driven design flow

When there are timing constraints, the design flow becomes a constrained optimization, which is called a timing-driven design flow. As shown in Figure 2.15, upon the completion of each stage, the satisfaction of the timing constraints is tested. Looping is necessary if timing violations are identified. The main cause of these iterations is wiring effects. As mentioned before, a significant part of the path delays are wire delays. We have also stated that the wire topologies might not be finalized until the completion of the routing. These make accurate computation of delays and evaluation of constraint satisfaction very difficult during logic synthesis. Even during placement, wiring still needs to be estimated. Estimation errors may lead to timing problems, and the problems may occur very late in the design flow. Another cause of this non-convergence is that some algorithms do not directly minimize delays. For instance, many placement algorithms focus on minimizing the total wire length, which is helpful in reducing delay but cannot guarantee it.

A timing-driven design flow will be given in more detail in Chapter 5.

2.5. Challenges in the Deep Sub-Micron Era

IC fabrication technology is often labeled with the minimum feature size, that is, the minimum width of the patterns that can be correctly fabricated. The feature size has been shrinking from 1-micron (about ten years ago) to about 0.1-micron (sub-micron, today). And it is expected that the feature size will reach 0.05-micron (deep sub-micron) in a few years.

2.5.1. The Deep Sub-Micron Challenges

Smaller feature sizes mean higher density and shorter intrinsic gate delays. However, all the parameters are not scaled down at the same rate. Unfortunately, parameters like wire resistance and wire capacitance are scaled down slowly, such that wire delay becomes a dominant part in the total delay. Other new problems such as manufacturability also emerge. We discuss three DSM problems: timing closure, manufacturability and library migration.

2.5.1.1. Timing closure

The timing closure problem arises from the fact that the design flow contains a set of sequential steps. Early steps have to predict what the later steps will do. Inaccurate prediction may lead to wrong decisions, which can only be recognized later. When this happens, design iteration is necessary. For instance, at the synthesis stage, the wire delays are not known. So the synthesizer has to estimate wire length based on probability or some other models. When the synthesized circuit is passed to the physical design tool, it might become clear that some wire lengths were incorrectly estimated.

Unfortunately this is not the end of the nightmare, because using the "exact" wire length just derived from the routed design to control a new round of synthesis does not guarantee the same layout, and thus does not guarantee those wire lengths. In fact timing closure has been a problem for a long time, but DSM promotes it to the top of the problems.

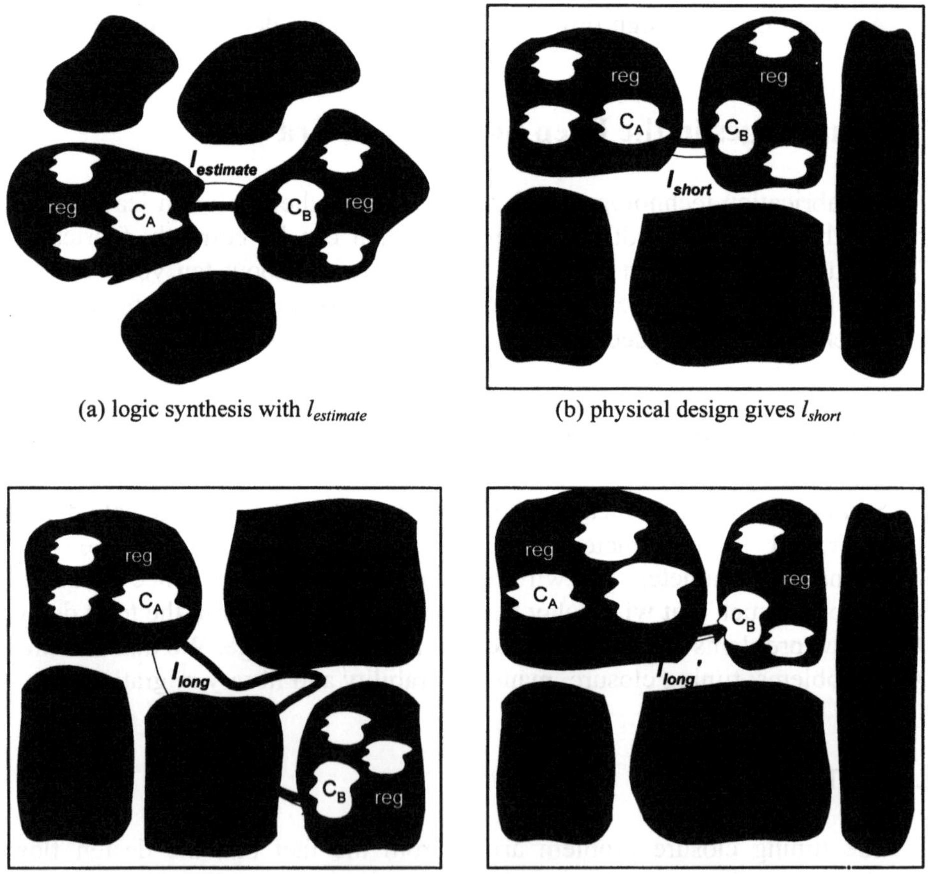

(a) logic synthesis with $l_{estimate}$

(b) physical design gives l_{short}

(c) physical design gives l_{long}

(d) physical design gives l_{long}'

Figure 2.16: An example of the timing closure problem

An example of the timing closure problem is shown in Figure 2.16. A path bridging two modules is highlighted. C_A and C_B are the part of the combinational logic that relates to the path. The two modules also contain logic that is irrelevant to the path. Figure 2.16(a) is the logic synthesis stage, in which the length of the wire under consideration can only be estimated, denoted by $l_{estimate}$. Logic synthesis takes the estimated wire length and distributes the path delay among C_A, the wire and C_B. Physical design may result in a real wire with length l_{short} that is very close to $l_{estimate}$, as shown in Figure 2.16(b); thus the previous estimation is legal and timing closure is achieved. However, physical design may result in a layout shown in Figure 2.16(c), in which the two modules are placed apart such that the wire length l_{long} deviates substantially from $l_{estimate}$. This shows that the separation of logic synthesis and physical design is a source of wrong estimation. Another source is the logic synthesis itself, because the module size is a result of the synthesis, and thus a result of the estimated wire length. In Figure 2.16(d), the placement seems to be fine, because the modules carrying C_A and C_B are close to each other. However the module carrying C_A becomes so large that the wire length l_{long}' still becomes illegal.

The above discussion only involves one path. In SOC designs there are many paths. Also ignored in the discussion is the routability of the wires. Sometimes wires have to detour to fulfill routability, such that the wire lengths and delays become even harder to predict at the early stages like logic synthesis.

To illustrate another possible cause of non-convergence, an example is given in Figure 2.17. We focus on the module carrying combinational part C_A. The synthesis, as shown in Figure 2.17(a), works with an estimated wire length $l_{estimate}$. After physical design, as shown in Figure 2.17(b), it turns out that the wire length l_{long} is too long such that the path constraint is not met. A new iteration is called with $l_{estimate}$ replaced by l_{long}. The synthesis now deals with a tighter timing requirement, because C_A needs to be faster to compensate the larger wire delay. As shown in Figure 2.17(c), this results in larger area of the module, mostly due to the increased size of C_A. When the new synthesis result is passed to physical design while the original relative module positions in the layout are maintained, as shown in Figure 2.17(d). The previous problem goes away, if the new wire length l_{long}* is consistent with l_{long}. However a new problem may occur. Consider a previously good wire, the one with length l_{other} in Figure 2.17(b). Due to the increase size of the re-synthesized module, this innocent wire may have its length increased to l_{other}*, causing a new timing problem.

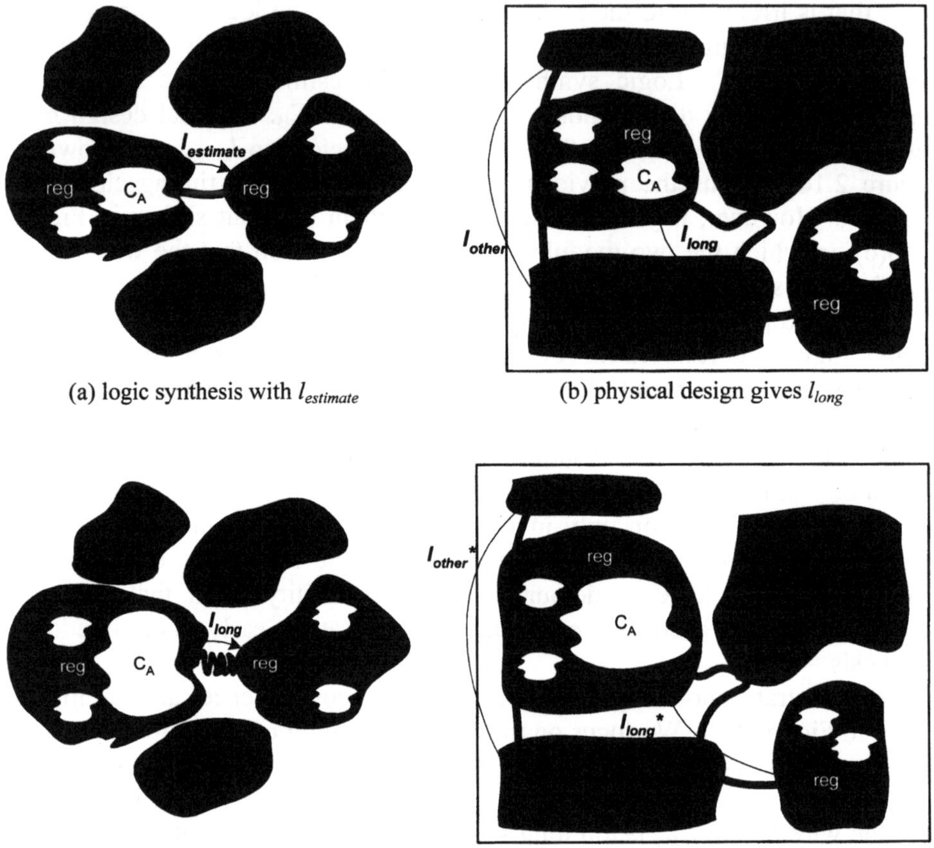

(a) logic synthesis with $l_{estimate}$ (b) physical design gives l_{long}

(c) logic synthesis with l_{long} (d) physical design gives l_{long}*

Figure 2.17: Another example of the timing closure problem

2.5.1.2. Manufacturability

As the limits of the mask-making system (finite aperture of projection) are reached with smaller geometries, the actual layout patterns on the wafer differ from those produced by a CAD tool [PLUM00]. The patterns after the lithography may have unexpected shapes such as collapsed edges and rounded corners. Although pre-distortion can be added to offset some of the real distortions, the number of layout patterns generated by a conventional design flow can make this task take an unreasonable amount of time and generate an enormous data set.

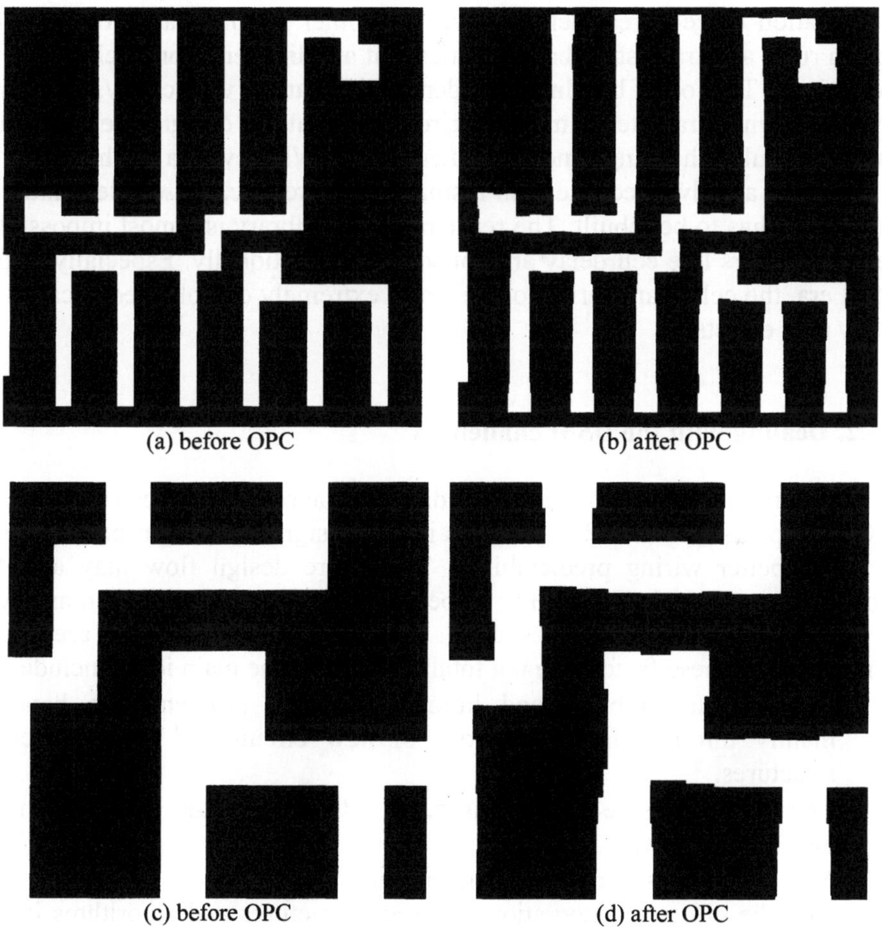

Figure 2.18: The OPC process.

Figure 2.18 illustrates a simple example of the Optical Proximity Correction (OPC). Design-For-Manufacturability (DFM) has become one of today's hot research topics.

2.5.1.3. Library migration

Standard-cell is the most widely used circuit structure in IC design, which will be detailed in Chapter 3. This structure involves a library that contains several hundred library cells with various logic functions and driving capacities. Each cell needs to be manually designed to get an optimal combination of cell area, delay, power and other factors, based on a set of design rules and transistor characteristics that are different from technology to technology. The job of building a modern cell library is very costly, involving layout design, parameter extraction, circuit simulation, design rule checking, electrical rule checking and documentation. Whenever a technology is migrated to an advanced one with a smaller feature size, the whole standard-cell library has to be rebuilt. The reuse of the old library is almost impossible, because factors like cell delay are not scaled proportionally. Especially in the DSM era, the cell characterization becomes extremely complicated because of many new effects.

2.5.2. Dealing with the DSM challenges

To deal with the DSM challenges discussed above, efforts can be made in several aspects. Individual algorithms in the design flow might be improved to allow better wiring predictability. The entire design flow may also be modified to reduce the possibility of being trapped in endless iterations. New circuit structures are also sought, which offer predictable area and performance. These factors are not totally isolated. The main ideas include:

- The manufacturability and library migration problems are handled mainly through the applications of new circuit and interconnection structures.
- The timing closure problem is dealt with through the development of new design flow.
- New algorithms bridge the structures and the flow.

We study the effective integration of circuit structures and algorithms into a design flow. This does not mean that new structures and algorithms that do not fit in the new design flow are worthless; they are still valuable.

It has been proposed [MO02a] that regular circuit structures, sometimes called "regular fabrics", can provide better guarantees that the layout designed by a CAD tool is correctly replicated in the fabrication. Regular fabrics use extensively simple and repeatable layout patterns in building the circuits. Memory and Programmable Logic Array are two typical regular structures. A

PLA is composed of regular patterns and its area and delay are directly related to the logic function it implements. A network of Sum-of-Products (SOP) nodes can be mapped directly to a PLA [RUDE88]. Technology mapping, required in standard-cell designs, is not necessary, nor are placement and routing necessary for a single-PLA circuit. The PLA structure is also library free. PLA and its variations will be detailed in Chapter 3.

Regularity exists not only in the lower layers of the layout, where transistors are made, but also in the interconnection layers. Regularity in the interconnections is very helpful in predicting wiring effects. A new regular global wiring scheme will be discussed in Chapter 4.

Regular structures certainly offer simpler and possibly better optical correction in enhancing manufacturability. Beyond what optical pre-distortion could do, regular fabrics can reduce manufacturing variations further. Unfortunately a quantitative measurement of any improved manufacturability of a layout due to regularity is still missing. The numbers of layout patterns, wire segments and vias might give can be reasonable metrics when comparing two designs in terms of manufacturability, but this study remains to be done. When such a study becomes available, manufacturability can become another trade-off that can be made along with area, delay, power etc, and the role of regularity towards these goals can be assessed.

The separation of the logic synthesis and physical design in the traditional design flow is a major cause of timing closure problems. Following are two approaches that help solve the timing closure problems: (1) One approach is to do some physical design before synthesis starts, such that wiring effects can be more precisely estimated based on the pre-synthesis physical design results, for example the rough positions of the modules. Then logic synthesis uses these to guide its operations. If the synthesis results, module area for instance, have no conflicts with the pre-synthesis results, then convergence is reached. Today's commonly adopted "Physical Synthesis" flow is such an approach. (2) Another approach is to leave certain flexibilities in the synthesis results such that they can be used to solve potential timing problem during the physical design. The so-called Module-Based design flow that we have developed belongs to this category. In Chapter 5, these two kinds of design flows will be described and compared.

2.6. Summary

In this chapter, we presented the preliminaries of modern IC design, including CMOS transistor, static and dynamic configurations, delay modeling of gate and wire, various design styles and timing-driven design flow. New DSM challenges were discussed also. The strategies proposed in Section 2.5 to handle the DSM problems are the main topic in the following chapters.

Chapter 3

CIRCUIT STRUCTURES

3.1. Overview

The following circuit structures as well as their design methodologies are described in this chapter.

- Standard-cell (SC): An existing structure with a new integrated placement and routing algorithm.
- Programmable Logic Array (PLA): Review of an existing structure.
- Network of PLAs (NPLA): Review of an existing structure.
- River PLA (RPLA): A new structure and its design methodology.
- Whirlpool PLA (WPLA): A new structure and its design methodology.
- Checkerboard (CB): A new structure and its design methodology.

Structures like standard-cells and PLAs are not new, but worth having a quick review because of their importance in Very Large Scale Integrated-Circuit (VLSI) design. A novel physical design algorithm for SC is presented. River PLA, Whirlpool PLA and Checkerboard are relatively new; both their circuit structures and design methodologies are discussed. Since the proposed design flow is module based, we focus on the design of modules with these structures. The sizes of the modules normally range from several hundreds to ten thousand gates. Actually some of the structures such as standard-cell, NPLA and Checkerboard can be used for full chip design. The PLA based structures can be extended to programmable or reconfigurable versions, at a cost of some area and delay.

In this chapter, we only consider how to use a structure to implement a given netlist. The created layout is a module. To implement the whole chip, many such modules might be needed. The way such modules, possibly made of with different structures, are organized will be discussed in Chapters 5.

3.2. Standard-Cell

Standard-cell is the most widely used style in VLSI design. A standard-cell library is associated with a specific technology, containing tens to hundreds of library cells. Each library cell in the library implements a logic function. Several library cells may implement the same function, but they provide different driving capacities. A standard-cell netlist is composed of the instantiations of the library-cells (simply called "cells"). The term "standard" comes from the fact that all the library cells of a library have the same height (Y-size of the layout). When placing the cells in a layout, they can be arranged in rows.

3.2.1. The structure and usage of the standard-cell

Figure 3.1 shows a typical standard-cell (three-input NOR). It is shown in the layout view that a power stripe (vdd) is laid out horizontally at the top of the cell and a ground stripe is at the bottom. The PMOS transistors in the schematic view are placed in the upper half of the cell layout and the NMOS transistors in the lower half. The input pins, which are normally placed halfway between the power and ground stripes, connect to vertical poly-silicon lines that control the PMOS and NMOS transistors. Most library cells occupy the poly-silicon layer and the metal 1 layer, and all the pins are on metal 1. The power and ground stripes are also on metal 1. The width of a standard-cell is mainly determined by the complexity of the schematic and the driving capacity. Simple and weak cells are usually narrow.

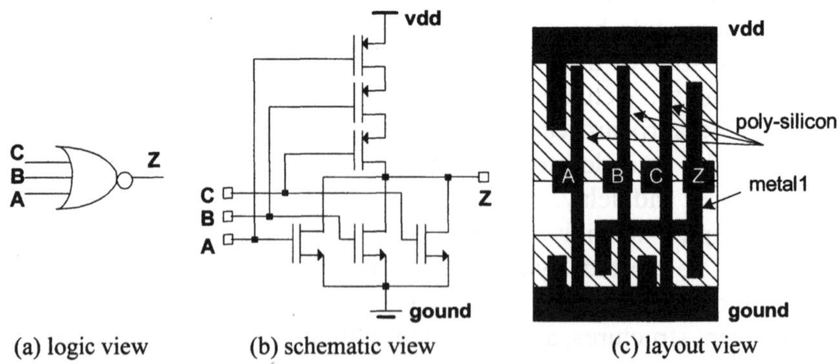

Figure 3.1: A typical standard-cell (three-input NOR).

The usage of the standard-cell is shown in Figure 3.2. In Figure 3.2(a), cells are placed in a rectangular region which is sliced horizontally into rows. The height of a row is the same as the height of the library cells. In the early days of the standard-cell design, as shown in Figure 3.2(a), the rows were split by channels in which wires are routed to connect the cells. The cells in a row may not be seamlessly placed, and they can only be X-flipped. However to supply power to all the cells in the row, a power stripe, which is on metal 1, extends along the entire row. So does the ground stripe. In this sense, the power and ground stripes on the cells can be regarded as power and ground pins. Power stripes of all the rows are connected on the left and/or right side of the rows; so are the ground stripes. As more metal layers become available for routing, rows can be vertically abutted and routing only takes place above the cells. Note that to abut two adjacent rows, the cells in one row must all be Y-flipped such that their power (or ground) pins are compatible with the power (or ground) of the abutted row. A simple approach to this is to let all even numbered rows, thus the cells placed in these rows, be Y-flipped. Modern physical design tools even allow cells to be placed in a non-rectangular region and obstructions like hard modules may exist in the region, as shown in Figure 3.2(c). The whole layout can be rotated, as shown in Figure 3.2(d), such that rows look like columns. As a convention, we call the columns "vertical rows", and the normal layout configuration "horizontal rows".

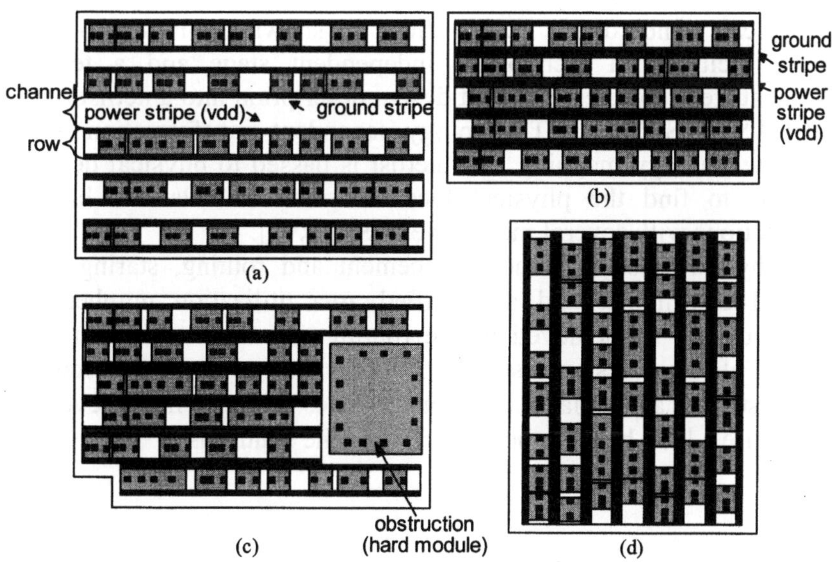

Figure 3.2: The usage of the standard-cells.

Finally the term "area utilization" is defined. It is the ratio of pure cell area and the layout area. The utilization is no greater than 1. Even in the abutted-row case, the area utilization cannot get very close to 1 for two reasons. One is that the widths of the cells are fixed. When cells are placed in a row, they may not fully occupy the row. But of course one row in the layout may have all cells in it placed without white space, and that row determines the row width of the layout. The other rows may not have 100% utilization. The other reason is that sometimes gaps must be created between the cells just to increase the layout area such that more routing resources are allocated. Therefore, especially in the abutted-row case, neither layout area nor utilization is a direct goal.

The standard-cell design flow includes three stages. The first stage, called logic synthesis, generates a netlist of cells which instantiate library cells. The next two stages, usually called the physical design, is to place the cells in rows (placement) and then to make the interconnections of the cells (routing). The existing design methodology for standard-cell is discussed in the next subsection. After that, we will present a new physical design algorithm for standard-cell, in which placement and routing are integrated.

3.2.2. Existing design methodology for the standard-cell

A typical standard-cell design flow is shown in Figure 3.3. Logic synthesis, including a technology independent stage and a technology dependent stage, translates the input design description into a netlist of library cells [BRAY90, SENT92]. Technology dependent transformation is widely known as technology mapping. The netlist is passed to physical design; and the goal is to find the physical locations for the cells and build their interconnections with several metal layers [SHER93].

The physical design includes placement and routing, staring with the synthesized cell netlist and a specified area utilization number. If full routability cannot be achieved in the routing, the area utilization must be altered and physical design restarted. Another kind of iteration occurs when timing constraints are violated. Its consequences are even more severe, since it may require a loop back to the beginning of the synthesis.

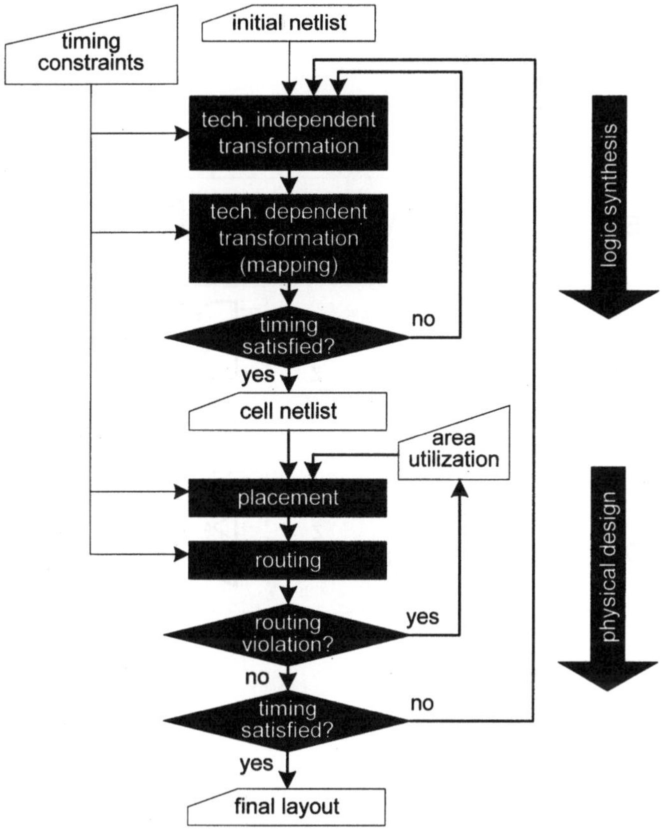

Figure 3.3: The standard-cell design flow.

Following are overviews of the standard-cell synthesis and physical design stages. Physical design is given in some detail, because in the next subsection a new standard-cell physical design algorithm will be presented.

3.2.2.1. Synthesis of the standard-cell design

Logic synthesis involves a technology independent stage and a technology dependent stage.

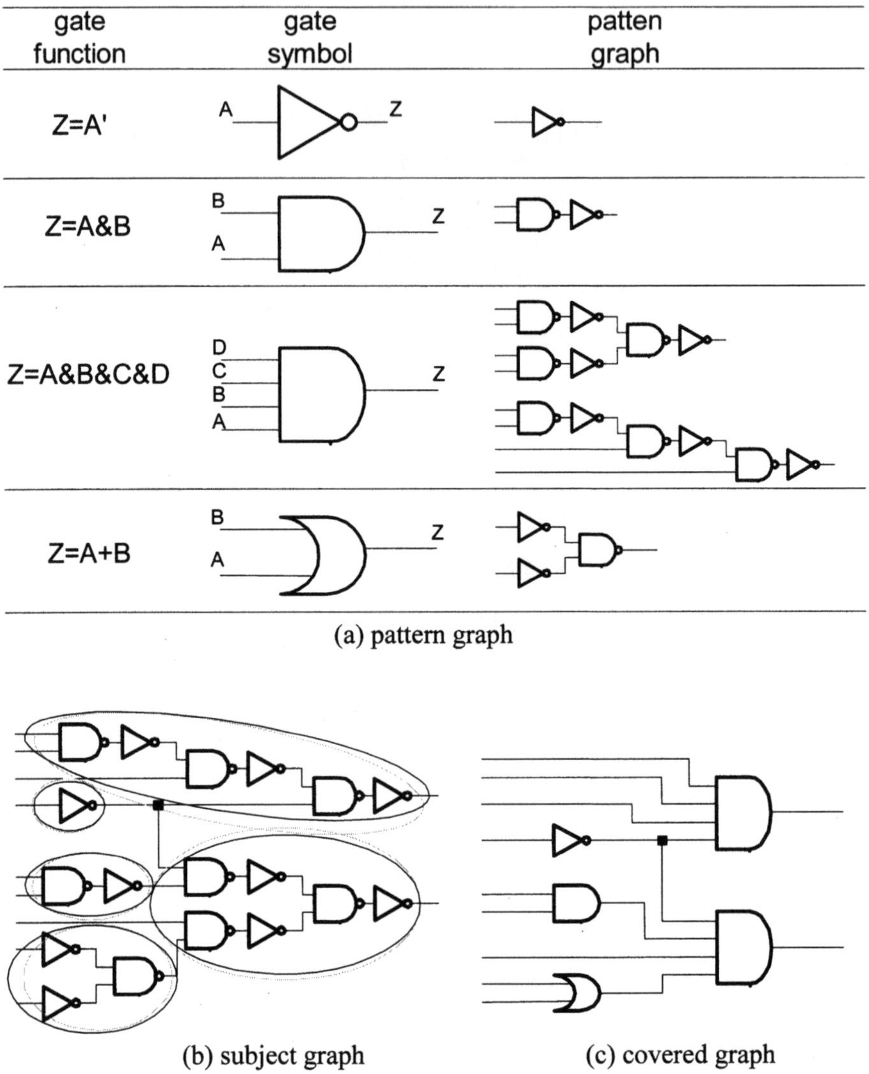

(a) pattern graph

(b) subject graph

(c) covered graph

Figure 3.4: An example of the technology mapping.

In the technology independent stage, the initial logic network is transformed via Boolean operations. The approach identifies and eliminates the redundancies in the logic functions and merges common sub-expressions [BRAY90]. The second stage, usually called technology mapping, transforms the Boolean network to a netlist of the standard-cells.

An example of technology mapping is shown in Figure 3.4. The network produced in the technology independent stage can be further decomposed into a network of NANDs and INVs, called a subject graph as illustrated in Figure 3.4(b). Each library cell in the standard-cell library is viewed as pattern graph(s), because the logic function of a library cell can also be represented as an interconnection of NANDs and INVs. Figure 3.4(a) shows a library of four library-cells and their pattern graphs. Notice that the library-cell "Z=A&B&C&D" corresponds to two pattern graphs with the same logic function but different topologies. The technology mapping problem is to cover the subject graph with the pattern graphs, or to instantiate library-cells to implement a netlist that is logically equivalent to the original Boolean network. The covered graph of the circuit in Figure 3.4(b) is shown in Figure 3.4(c). A library cell has an area cost and a delay cost. To allow more design freedom, there can be several cells that represent the same logic function, but they have different areas and delays. Such cells correspond to discrete points on an area/delay trade-off curve of the logic function. Technology mapping tools have the responsibility to pick up the appropriate cells such that some combination of total area and/or delay is reduced.

3.2.2.2. Physical design of the standard-cell

After the logic network is mapped to a network of standard-cells, physical design starts. Physical design consists of two steps, placement and routing. Standard-alone SC placers and routers are commercially available, and usually there is no restriction on pairing a specific placer with a specific router. Placement and routing are two separate steps in most existing standard-cell physical design flows.

The goals of a SC placer include, besides the non-overlapping (cells cannot overlap) requirement: (1) reduce the layout size unless the area utilization is specified; and (2) make the placement highly routable. The two goals usually conflict. Since the total cell area is a constant, reducing layout size alone is not hard. However routing congestion may require that white space be placed between cells. On the other hand, routability alone is easy to ensure if cells are placed sparsely enough. The problem is that at the placement stage, it is hard to determine how much white space is to be kept and where [YANG02]. At the placement stage, routability needs to be estimated. Estimations are based on certain net models, or the behavior of how nets are routed by the router. This is not easy, especially when the placer and the router are from different sources. Run time, a practical issue, prevents us from calling the real router frequently during the placement. Thus, simplified net models have to be used in the placement. Minimum Rectilinear Steiner Tree (MRST), a commonly used net model in the routing, is too expensive in computation to be adopted in the placement. Although the solution to a MRST problem is NP-complete, some heuristics provide near-optimal solutions while the run times are short. Even if each net is routed as MRST, it is still possible that different nets may compete for routing resources, that is, the metal wires. This leads to the routability problem. Therefore approximation models are necessary. A simple and efficient one is the half perimeter model, as described in Section 2.2.4.3.

Several kinds of standard-cell placement algorithms are available. One is successive bi-partitioning, which minimizes the cut number (number of wires crossing the cut) at each partitioning step [SHER93, WANG00, OU00]. Another is the so-called force-directed method, which simulates the mechanical problem of particles connected by springs [KLEI91, VYGE97, QUIN79, EISE98]. In addition, simulated-annealing based algorithms also work well in solving the standard-cell placement problem [SECH88]. After placement is done, the cell positions and the layout area and shape are fixed. The goal of the routing is to construct the interconnections of the cells without any geometric violations (100% routable).

If finally routability turns out to be a problem, iteration back to placement is needed and usually the area utilization number is reduced to produce more routing resources. Wrong estimation of routability in the placement is the main reason for iterations. Even in the early phase of the routing, called global routing, the detailed routing behavior is estimated and possible congestion identified and removed. The global router just uses a more accurate net model, and of course is more time consuming than that used in placement. Wrong estimation in placement and even global routing may cause failure of the detailed routing, the final phase of routing.

3.2.3. An Integrated Standard-cell Physical Design

3.2.3.1. Overview

Various approaches have been sought to close the non-convergence in the SC physical design stage. A common way is to try to make the estimation in the placement more accurate [WANG00, YANG02, YILD01, KAST00, CALD00]. Another way is to allow the router to locally modify the placement and/or reroute to fix the routing violations [GROE89, TZEN88, SHIR96]. All methods still separate placement and routing.

An Integrated Standard-Cell Physical Design (ISCPD) algorithm is proposed, which integrates the standard-cell placement and routing into a single step. Figure 3.5 gives the new standard-cell design flow with ISCPD replacing the separated placement and routing. ISCPD guarantees 100% routability, because it adjusts area utilization automatically.

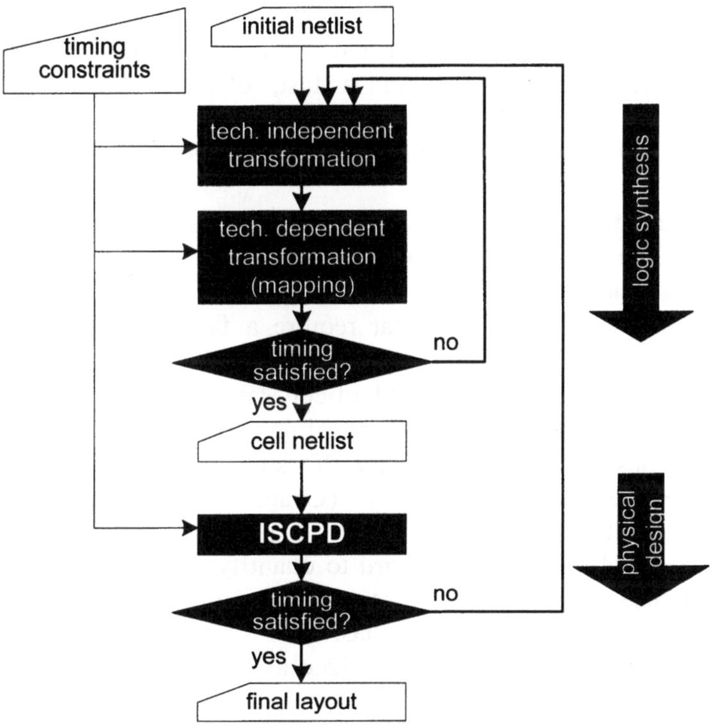

Figure 3.5: The ISCPD in the standard-cell design flow.

A two-layer (the two upper layers, metal 2 and 3) spine net topology is employed. The bottom layer (metal 1) is used to make connections between the standard-cell pins and the spine nets. The spine of a net is a horizontal wire segment on metal 3 connecting the SC output pin driving the net. SC input pins of the same net are connected to the spine using vertical segments on metal 2 called "ribs". Given sequences of cells in the rows, the cell placement and arrangement of the ribs of the nets are done from the left side of the layout towards the right, followed by the arrangement of the spines of the nets. An example of the spine topology can be found in Figure 3.7. The simple spine net topology makes the construction of the interconnections fast. The sequences of the cells in the rows uniquely determine the layout. The initial sequences are generated through recursive bi-partitioning. The partitioning benefits from the fixed spine net topology, making the term "cut" more meaningful. Simulated-annealing can be used as a final step to search for possible improvement by randomly moving or swapping cells in the sequences and swapping pairs of I/O ports. Experiments have shown that the layout generated immediately after the partitioning is already close to the final layout done through annealing in terms of area and wire length. The partitioning part of the algorithm is fast, offering a means of quick construction of a 100% valid standard-cell layout with acceptable quality.

ISCPD is attractive to the timing-driven design flow, detailed in Chapter 5, for the following reasons:

(1) The integrated placement and routing prevents complex but inaccurate routing estimation in the placement.

(2) Although the ISCPD does not directly aim at timing closure, it helps algorithms that do so and that require a fast but good standard-cell physical design tool. The partitioning part of ISCPD provides a quick approach to obtaining a standard-cell layout with good quality. To pursue even higher quality, the annealing part of the ISCPD can improve this in a post-processing stage. ISCPD is the physical design tool used in our Module-Based design flow (Chapter 5) for the generation of standard-cell modules.

(3) Manufacturability, although hard to quantify, is enhanced indirectly by making use of the simple and regular spine topology, which reduces the numbers of wire segments and vias.

The rest of the discussion is organized as follows. Section 3.2.3.2 gives the preliminaries of the cell model and spine net topology. The ISCPD algorithm is described in Section 3.2.3.3. In Section 3.2.3.4, some practical issues are discussed. Experimental results are given in Section 3.2.3.5, and Section 3.2.3.6 concludes.

3.2.3.2. Preliminaries

The metal layers are denoted by M1, M2 and M3. The geometries on a metal layer are restricted by the design rules. In this algorithm, we consider grid routing and each layer uses a uniform wire pitch. Denote the wire pitches on M1, M2 and M3 by *Pitch*(1), *Pitch*(2) and *Pitch*(3). Adjacent layers are connected by vias; the vias are also on grid. A series of grids on the same layer with the same X (or Y) coordinate is called a "routing track". Upper-case *X*/*Y* variables represent the global or absolute distances on the layout, while lower-case *x*/*y* variables represent local distances in grid units. The subscripts of the variables usually indicate the layers on which the distances are measured, e.g. X_{M1}, y_{M2}.

A cell is a rectangle with size of $W(c)$ by H_C. All the cells have the same height H_C. The pins of a cell are all on M1; other parts of M1 are occupied by the internal connections of the cell. A pin can be reached either from its surrounding region on M1, if not blocked by existing M1 geometries, or from M2 by a via.

Figure 3.6(a) is a typical layout of a NOR3 gate. It is abstracted in Figure 3.6(b), in which hatched regions represent the M1 areas already occupied by power, ground and other internal connections, and the dotted region is the M1 area occupied by the output pin Z. The remaining white space is free for M1 routing. The ISCPD algorithm only uses part of the free M1 space, so the layout is further adjusted to Figure 3.6(c), in which the M1 blockages are expanded, and the output pin and input pins are isolated. It is also assumed that all the input pins have the same Y-grid, and parameters $l_{B\text{-}IP}$ and $l_{A\text{-}IP}$ denote the number of free M1 horizontal tracks below and above the pins. For the output pin, $l_{L\text{-}OP}$ denotes the number of vertical M1 tracks to its left. The majority of the cells in a real library can be translated to their adjusted views without modification of their layouts. A few cells need slight modification with almost no area and delay overhead. Only single-output cells are handled, but extension to cells with multiple output pins is easy.

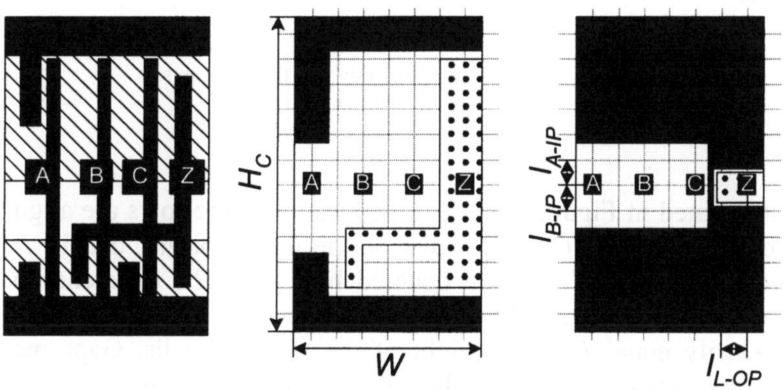

(a) layout (b) abstract view (c) adjusted view

Figure 3.6: The typical layout of a cell and its views.

The cells are placed in horizontal rows, labeled $0,1,..,N_R-1$, where N_R is the number of rows. All the odd rows are vertically flipped. If two adjacent rows are seamlessly stacked without a routing channel between them, either their power or ground stripes touch. The "sequence" of row r, denoted by $S(r)$, is an ordered list of the cells, from left to right, that are placed in the row.

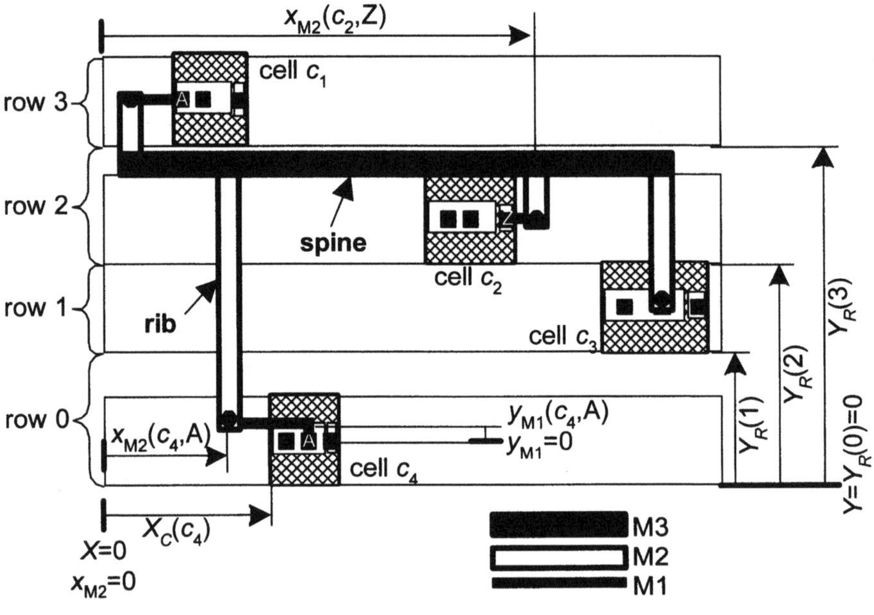

Figure 3.7: The spine topology and the notations.

As illustrated in Figure 3.7, the left edges of all the rows are aligned. Let the bottom-left corner of the 0-th row be the global origin, that is, X=0 and Y=0. Each row has a Y-position denoted by $Y_R(r)$. The difference between $Y_R(r+1)$ and $Y_R(r)$ is the height of row r. In general, the height of a row may not necessarily equal H_C, the height of the standard-cells. Gaps may exist between two rows to accommodate horizontal wires, increasing the row height. The gap between row 2 and 3 in Figure 3.7 is an example.

The spine is a horizontal M3 track in the row of the output pin driving the net. All the input pins of the net connect to the spine by vertical M2 ribs. The spine topology of net n is shown in Figure 3.7, connecting the output pin of cell c_2 and input pins of cells c_1, c_3 and c_4. (A connection to an input pin, except the left-most one, first goes "above" or "below" on M1, and then horizontally goes to the left of the input pin on a M1 track with Y-grid of $y_{M1}(cell,ipin)$. The left-most input pin connection always directly goes left. The horizontal M1 track ends at $x_{M2}(cell,ipin)$, the X-grid of the M2 rib. An output pin connects horizontally to $x_{M2}(cell,opin)$, going up through a via to M2. A vertical M2 track spans to below the spine and goes up to the spine through another via.

3.2.3.3. The integrated placement and routing

The ISCPD algorithm is modular. The first part is a recursive bi-partitioning step, from which the sequences of cells in the rows, $S(r)$'s, are produced. A procedure called "All_Cells()" builds the layout from the $S(r)$ information. Further improvement is possible at the expense of longer run time; a simulated-annealing framework searches for a better solution by swapping or moving cell(s) in the $S(r)$ sequences or swapping I/O ports of the SC module. Each random move of simulated-annealing invokes a run of All_Cells(). The annealing can stop any time the user likes, and still the layout is guaranteed to be fully routed.

In 3.2.3.3.1, a procedure called One_Cell(c) is described that instantiates a cell into a row and creates the ribs for its input pins and the connection of its output pin. Procedure All_Cells(), discussed in 3.2.3.3.2, is a scheduler that selects a cell from a sequence and calls One_Cell(cell) to put it in the row. After all the cells are inserted in the rows, All_Cells() routes the spines row by row. I/O connections are a little different, and are discussed separately in 3.2.3.3.3. The bi-partitioning and simulated-annealing are discussed in Sections 3.2.3.3.4 and 3.2.3.3.5 respectively. Finally in 3.2.3.3.6, how to determine the row number is discussed, although this really takes place at the beginning of the algorithm.

3.2.3.3.1. One_Cell(*c*)

We discuss cell c in an even row r=2, which is not vertically flipped. For odd rows, which are vertically flipped, the algorithm is the same, except "above"/"below" and "up"/"down" are exchanged. Before c is placed, other cells may have already been placed and their ribs routed. Therefore for each row we create two variables, $x_{M2\text{-}MAX}(r)$ and $X_{MAX}(r)$, to represent the current end of all M2 tracks crossing the row and the right end position of standard-cells already placed in the row, respectively. These are the points from which the search is started towards the right for available M2 tracks to place ribs and for empty space to place the cell.

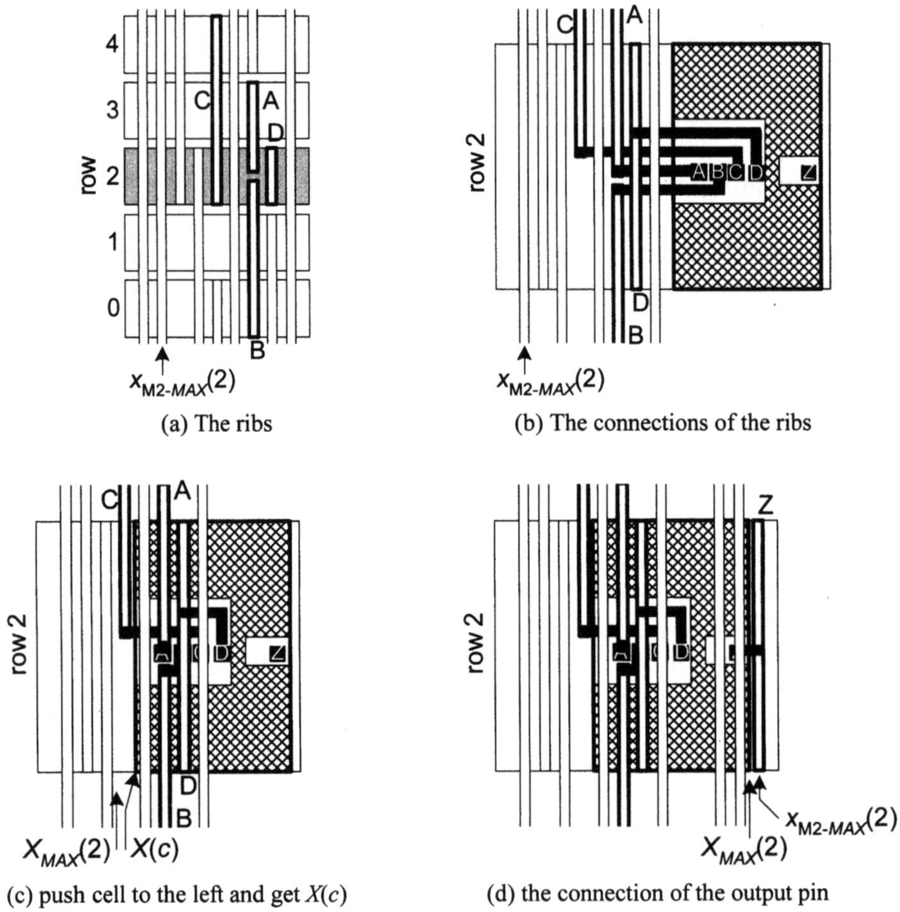

Figure 3.8: Connections of the input pins and placement of the cell.

An example is shown in Figure 3.8. Cell *c* has input pins A, B, C and D and output pin Z. The cell is being placed in row 2, and $x_{M2}(2)$ and $X_{ROW\text{-}MAX}(2)$ are the current end M2 track and right end position of row 2. To simplify the discussion, we say a rib being "up", "down" or "this", if the spine that the rib connects to is in an upper, lower or this row. Since the spines have not been arranged yet in the rows, the only information is the row locations of the spines. Therefore, we assume conservatively that a rib, "up" or "down" occupies the full vertical M2 track of the spine's row, as A,B and C in Figure 3.8(a). If the rib is "this", as D in Figure 3.8(a), its spine is in the current row. It then occupies one whole vertical M2 track of the row. Note that when the spines are finally routed, their Y-coordinates are known and the over-occupied M2 parts of the ribs can then be truncated to improve the layout.

Starting from $x_{M2}(2)$ towards the right, each vertical M2 track is examined to see whether it can be used by any rib(s) of the input pin(s). If not, proceed with the next M2 track. If it is usable to any rib(s), choose the longest unassigned rib to fill in this track. Then see if a remaining unassigned rib in an opposite direction can fit in the same track. Obviously rib of kind "this" cannot share an M2 track with any other rib. The first step, shown in Figure 3.8(b) is to arrange the ribs and build their connections to the corresponding input pins. A rib on the right side of its input pin is not allowed in ISCPD. An input pin routes on M1 to its rib. The leftmost pin just uses a horizontal M1 wire, and the rests may either go "above" or "below". If two pins both go "above" ("below"), the pin on the right must be above (below) the pin on the left. The arrangement of ribs is detailed later in this subsection, which results in $x_{M2}(cell,ipin)$'s and $y_{M1}(cell,ipin)$'s of the input pins. The cell is then pushed to the left, as illustrated in Figure 3.8(c), until either it touches $X_{MAX}(2)$, or any of the input pin reaches its vertical M2 rib. Thus the cell position $X_C(cell)$ is obtained, and the right edge of the cell becomes the new $X_{MAX}(2)$. Next, as shown in Figure 3.8(d), the output pin searches for its M2 track, leaving room for later connection to the spine on M3 in this row. The M2 track of the output pin is just as the "this" rib, in the sense that it occupies one full M2 track in this row and the track can not be shared by others. Recall that the output pin has a region around it in the cell (the dotted region in Figure 3.6(b)) such that only the output pin itself can use the M1 resource in this region. This means that the M2 track of the output pin becomes the new $x_{M2\text{-}MAX}$ of the row, because no rib of the next cell in this row can be placed on its left. The updating of $x_{M2\text{-}MAX}$ and X_{MAX} is done in All_Cell(), which actually employs a more complicated but better strategy in doing this. The values obtained here form the lower bounds.

The difficult part is the arrangement of the ribs and their connections from the input pins. Recall that in the adjusted view of Figure 3.6(c), parameters $l_{B\text{-}IP}$ and $l_{A\text{-}IP}$ were defined that limit the horizontal M1 tracks below and above the input pins. Although for almost all library cells $l_{B\text{-}IP}+l_{U\text{-}IP}$ is greater than or equal to $N_{IP}(c)-1$, the number of input pins of cell c excluding the leftmost one, $l_{B\text{-}IP}$ or $l_{A\text{-}IP}$ alone may be smaller than $N_{IP}(c)-1$. This prevents us from arbitrarily selecting y_{M1} for the input pins. A heuristic is that when a M1 track is assigned, we try to keep the numbers of the remaining "above" and "below" M1 tracks as balanced as possible. The other problem is the opportunity of allowing two ribs to share the same M2 track. Of course the sharing can only occur if the two ribs are valid ribs for that M2 track, and one of them "goes down" while the other "goes up".

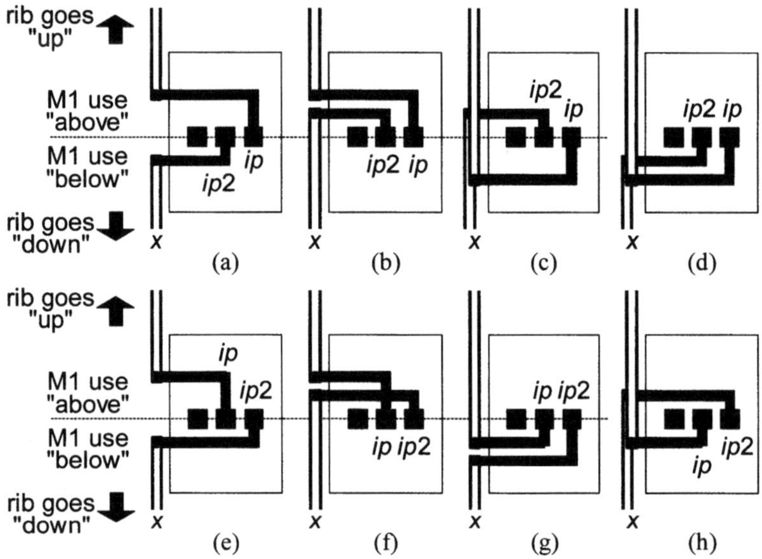

Figure 3.9: The M2_Sharing_Rule.

In Figure 3.9, the rule for two ribs sharing the same M2 track is illustrated. In the figure, *ip* is already assigned since it has the longest rib that fits in the M2 track at *x*, and the rib of *ip* goes "up". Now we test if another rib *ip*2 going "down" can share the same M2 track. The first case, Figure 3.9(a)-(d), is that *ip*2 is to the left of *ip*. If *ip* already "uses above", we only need to find a M1 track for *ip*2, either below or above, that is available, as shown in Figure 3.9(a) and (b). When doing the track selection, the balancing mechanism as described before is employed. If *ip* already "uses below", there will be no way to connect to the *ip*2 rib without causing overlap, as illustrated in Figure 3.9(c) and (d). The second case, Figure 3.9(e)-(h), is that *ip*2 is to the right of *ip*. If *ip* already uses "above", *ip*2 can only use "below" as in Figure 3.9(e). Crossing may occur if *ip*2 uses "above" as shown in Figure 3.9(f). If *ip* already uses "below", *ip*2 must also use "below", as in Figure 3.9(g). If *ip*2 uses "above" as in Figure 3.9(h), overlap occurs. Symmetrical analysis can be derived, in which *ip* goes "down" and *ip*2 goes "up". The rule also works for the leftmost input pin, but without distinguishing "below" and "above".

The following pseudo code summarizes the One_Cell(c) procedure, and it embeds the M1 balancing mechanism M1_Selection() and the M2_Sharing_Rule. An additional data structure *M2Occ*(x, r) records the occupation of the M2 track at x in row r. A value of 1 means the track has been occupied, and 0 unoccupied.

```
One_Cell(c):
/* Arrange the ribs for the input pins. */
belowCount=l_B-IP(c), upCount=l_A-IP(c), ipSet={all input pins of cell c}
for (x=x_M2(r); ipSet≠Φ; x++) {
   if (M2Occ(x,r)==1), continue.
   /* Collect input pins whose ribs can fit in the current M2 track. */
   vpSet=Φ
   for (all unassigned input pin ip) {
      rSp=the row of the spine of ip.
      if (M2Occ(x,r')==0, r'=rows the rib of ip covers), vpSet∪={ip}.
   }
   /* Choose the longest one to fit in the current M2 track. */
   if (vpSet==Φ), continue.
   Choose ip∈vpSet that has the longest |r−rSp|.
   vpSet−=ip, ipSet−=ip. M2Occ(x,r')=1,r'=rows the rib of ip covers.
   M1_Selection(ip, belowCount, aboveCount).
   /* Check if any other rib can share the same M2 track. */
   if (vpSet==Φ), continue.
   if (ip2∈vpSet that satisfies M2_Sharing_Rule) {
        vpSet−=ip2, ipSet−=ip2.
        M2Occ(x,r')=1,r'=rows the rib of ip2 covers.
   }
}
/* Determine y_M1 for the input pins. */
y_M1(leftmost pin)=0, yA=1, yB=1.
for (ip of the rest input pins from left to right) {
   if (go "above") { y_M1(ip)=yA++ } else { y_M1(ip)=yB-- }
}
/* Determine cell position. */
Push the cell towards left, until either it hits X_MAX(r) or
any input pin hits its rib. Obtain X_C(c).
/* Determine the connection of the output pin. */
for (x=x(opin)−l_L-OP;(opin); M2Occ(x,r)==1; x++).
x_M2(opin)=x, M2Occ(x,r)=1.
```

```
M1_Selection(ipin, belowCount, aboveCount):
if (ipin is NOT the left-most pin of the cell) {
    if (aboveCount>belowCount) { ipin uses "above", aboveCount--. }
                          else { ipin uses "below", belowCount--. }
}
```

3.2.3.3.2. All_Cells()

Given the sequences of the cells in the rows, $S(r)$'s, the rows of the spines are known. Starting from the left edge of an empty layout, a cell is picked up from one of the sequences and placed into its row by One_Cell(c). When all the cells are placed, the vertical M2 ribs have been constructed as well. Then the spines in each row are packed vertically by the left-edge algorithm [SHER93]. Following are the details of the algorithm; the major part is how to select the next sequence from which a cell is picked up and placed into the row.

An ordered list Q is set up, each entry being a row (a sequence). The rows are sorted in the ascending order of their current right end position $X_{MAX}(r)$. Each time the row at the head of the list is selected and its next unplaced cell c is handled with One_Cell(c). When c is placed and its ribs arranged, $X_{MAX}(r)$ and $x_{M2\text{-}MAX}(r)$ are updated. If the row still contains unplaced cells, it is inserted back into Q at the appropriate position according to its new $X_{MAX}(r)$. The method for updating $X_{MAX}(r)$ is important. The simplest way, already mentioned in the last sub-section, is to use the right edge of the cell just put in the row, $X_C(c)+W(c)$, as the new $X_{MAX}(r)$, and $x_{M2}(opin)$ as the new $x_{M2\text{-}MAX}(r)$. However this sometimes causes the rows at the top and bottom to finish cell filling earlier than the central rows. An explanation is that more vertical ribs pass through the center rows than for top and bottom rows. This results in cells in center rows not being placed close to the previous cells, in contrast to the cells in top and bottom rows. Experiments have demonstrated that the simple updating scheme usually leads to larger area and longer wire length. A better way is to take into account how other rows are currently filled. Variable $X_{MAX\text{-}ALL}$ keeps track of the maximum $X_{MAX}(r)$ among all rows. Every time cell c is filled into row r, the following four steps are carried out:

$$X_{MAX}(r) = X_C(c) + W(c)$$

$$X_{NAX-ALL} = \max\left(X_{MAX-ALL}, X_{MAX}(r)\right)$$

$$X_{NAX}(r) = \max\left(X_{MAX}(r), X_{MAX-ALL} - \xi\right)$$

$$x_{\mathrm{M2}-MAX}(r) = \max\left(x_{\mathrm{M2}}(c, opin), \frac{X_{MAX}(r)}{Pitch(2)} - \rho \right)$$

ξ in the third formula is a constant and its value is empirically chosen as $10Pitch(2)$. In the fourth formula, ρ is also an empirical constant with a value of 10. The idea is to keep all rows marching with similar paces.

When all the cells are placed and the ribs arranged, the left and right terminal positions of the spines are known. Also known are the rows to which the spines belong. Hence the remaining job is to arrange spines in each row, which determines the Y-coordinates of the spines. The spine arrangement adopts the "left-edge" algorithm, which is greedy but gives an optimum solution [SHER93] and was widely used in channel routing. Pick up an unassigned spine, which has the smallest X-coordinate of the left terminal and fits in the current horizontal track, and assign it to the track. When the current track can accommodate no more spines, a new track is created. The number of horizontal M3 tracks available within the height of a cell is $H_C/Pitch(3)$. If the required number of tracks exceeds this, a gap between the row and the row immediately above is created to produce extra tracks. But in the gap, M1 is also available; so

$$\frac{Pitch(1)}{Pitch(1) + Pitch(3)}$$

of the overflowing tracks are assigned to M3 in the gap and the remaining to M1. The spine packing starts from row 0. When a row is finished, its height is known, which may possibly include a gap to accommodate overflowing spine tracks, and used to set the Y-position of the next row. Finally the redundant M2 segments are trimmed, as mentioned in 3.2.3.3.1. The routing of the I/O ports is also done in All_Cells(), but will be clarified in 3.2.3.3.3.

After a run of All_Cells(), the X/Y sizes of the module is obtained. The All_Cells() procedure is summarized in the following pseudo code:

```
All_Cells():
initialize ordered list Q by putting in rows in arbitrary order.
X_MAX(r)=0, x_M2-MAX(r)=0 for r=0,..,N_R−1. XMAX=0.
while (Q≠Φ) {
  r=head of Q, c=next cell of S(r).
  if (c≠Φ) {
    One_Cell(c).
    update X_MAX-ALL, X_MAX(r) and x_M2-MAX(r).
    if (S(r)≠Φ), insert r in Q.
  }
}
Y_ROW(0)=0.
for (r=0; r≤N_R; r++) {
    use "left-edge" algorithm to pack spines.
    obtain row height H_ROW(r) and
    Y_ROW(r+1)=Y_ROW(r)+H_ROW(r).
}
Left/right I/O column packing (see subsection 3.4).
```

An example is shown in Figure 3.10, outlining the operation of All_Cells().

3.2.3.3.3. The I/O connections

Since the outline of the SC module is not fixed, the I/O ports are said to be "fixed" if their sequence on the boundary is fixed. User can give a fixed sequence of the I/O ports, or it can be generated randomly and improved by simulated-annealing.

There are four cases: input port (with output pin) on left/right, input port on top/bottom, output port (with input pin) on left/right and output port on top/bottom.

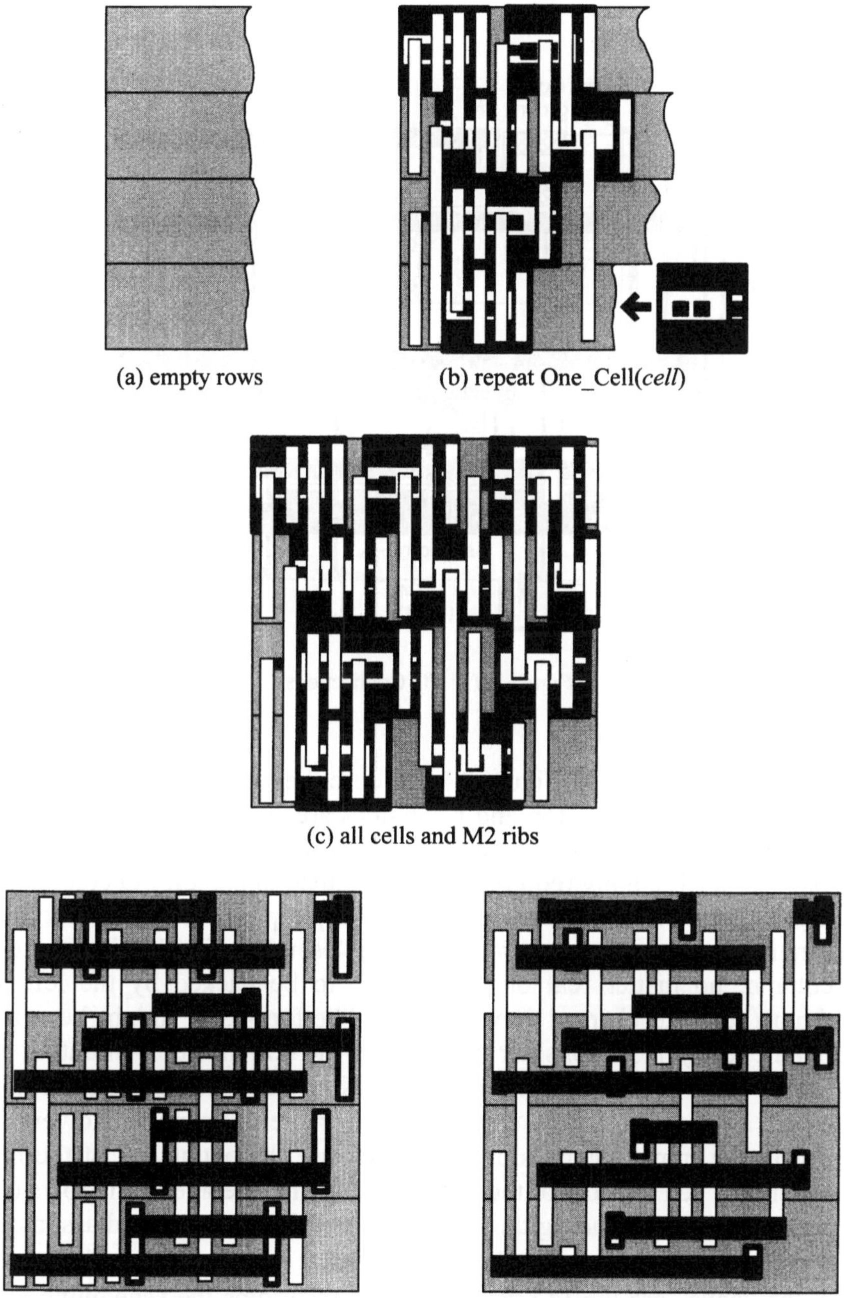

(a) empty rows

(b) repeat One_Cell(*cell*)

(c) all cells and M2 ribs

(d) packing of M3 spines in rows

(e) trimming redundant M2

Figure 3.10: The operation of All_Cells().

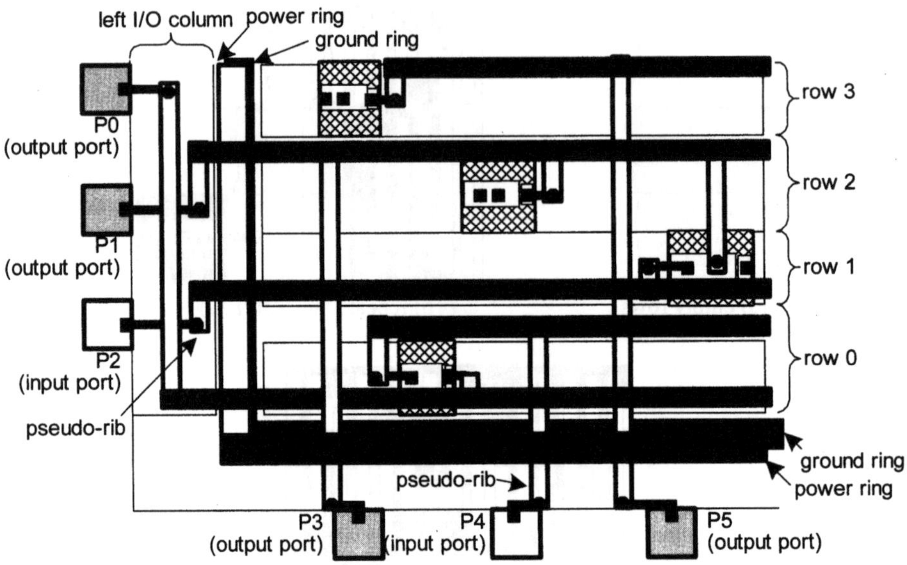

Figure 3.11: I/O Connections.

We use the example in Figure 3.11 to describe the I/O connections on the bottom and left sides. The top and right sides are symmetrical. First consider the ports on the left side. A so-called "left I/O column" is built. An output port, such as P0 and P1, simply requires its spine extend to the left I/O column such that it can connect to the spine with a rib. An input port, such as P2, creates its spine in the nearest row, through a so-called "pseudo-rib". The ribs in the left I/O column, no matter real or pseudo, are to be packed by the left-edge algorithm (here the channel is vertical, so "bottom-edge" might be a more appropriate term). Now consider the ports on the bottom side. An output port, such as P3 and P5, has a rib which connects to the corresponding spine. An input port, such as P4, uses a "pseudo-rib" to route to its spine. However the row where the spine occurs is unknown because the output pin of the net is not within any cell row but at the port. We take the average of the rows of all the pins on the net, including the output pin of this input port and assume that the port is in row −1. Imagine that there exists a "row −1" beneath row 0, in which the "cells" are the I/O ports. Row −1 also joins the scheduling of the cell rows in procedure All_Cells(). Its $X_{MAX}(-1)$ always takes the position of the next port to be handled.

Also shown in Figure 3.11 are the power/ground rings that surround the cell rows. They provide power/ground connections to the cell rows, and act as a mechanism of suppressing the latch-up effect. On the left/right sides, power uses M2; and it uses M3 on the bottom/top sides. Ground ring uses M1 only.

3.2.3.3.4. The generation of the initial placement — partitioning

The initial layout is generated through successive bi-partitioning and a run of All_Cells(). Fiduccia-Mattheyses (FM) partitioning algorithm is adopted [FIDU82] with a modified definition of "cut". In vertical partitioning, a spine crossing the cutting line counts as one cut. In horizontal partitioning, a rib crossing the cutting line counts as one cut. External pins with connections to pins in the current partitioning region can be considered if their relative positions to the current cutting line are known.

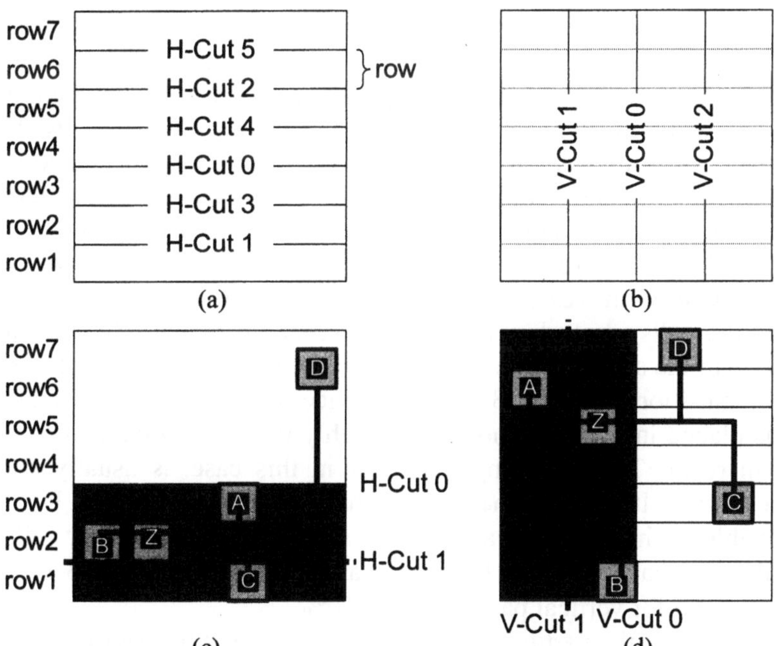

Figure 3.12: The bi-partitioning.

In Figure 3.12, the FM algorithm based on the spine model is illustrated. The sequence of the bi-partitioning is horizontal bi-partitioning and then vertical bi-partitioning, as shown in Figure 3.12(a) and (b), respectively. The partitioned region, the area that accommodates cells in each horizontal cut is an integer multiple of rows. The horizontal bi-partitioning stops only when single rows are reached. Vertical bi-partitioning is similar, except the stopping criterion is that a partition contains 5 or less cells (in a row). As partitioning progresses, the cells are restricted to smaller and smaller region. The objective of the bi-partitioning is to minimize the number of cuts.

The partitioning sequence given above guarantees that, in each cut, the pin distribution of a net across the cut is known. This feature is due to the spine model. Since the positions of the cells (carrying the pins of the net under consideration) with reference to the cut line are known, we can tell how the cut number of this net will be affected if a cell is moved from one side of the cut line to the other. A horizontal cut example is shown in Figure 3.12(c), in which the net under consideration contains output pin Z and input pins A to D. Before the current cut, some previous horizontal cuts have refined the Y-positions of some cells. This cut takes place in row1 and row2&3, the dotted region in the figure; so only cells carrying pin A, B, C and Z can switch between these two partitions. Suppose the cell carrying Z moves to row1, the spine also moves to that partition. The previous cut by the connection to pin C disappears, while the connections to A, B and D generate new cuts. Thus the total effect is an increase of 2 cuts. This shows the big difference between ISCPD partitioning and traditional FM. In FM, a net can only have zero cut or one cut, independent of net topology. However in the partitioning of ISCPD, the cut number has more realistic meaning, owning to the known net topology. If cell carrying C moves to row2&3, the cut of this net becomes zero. The move of the cells must obey some partition size balance; otherwise all cells may move to the same partition, which results in zero total cut. In the example, since the partition of row2&3 occupies twice area as row1, the total area of the cells placed in row2&3 should be roughly twice the total area of the cells placed in row1. The balancing ratio, 2/1 in this case, is usually allowed a small deviation. It is worth mentioning that the cell carrying pin D, although not movable in this cut, may affect the cut number if some other cells move. Vertical cuts have similar behavior, as shown in Figure 3.12(d). The balancing ratio for vertical partitioning is 50%.

I/O ports join the partitioning if they are not fixed; otherwise they only act as "cells" carrying external pins. The cells in each final partition (which may contain up to 5 cells) are randomly ordered. Thus the initial sequences of cells in the rows are generated. A run of All_Cells() produces the initial layout.

3.2.3.3.5. Simulated-annealing improvement

The initial layout generated by recursive bi-partitioning and one run of All_Cells() can be improved further by a simulated-annealing procedure. In the simulated-annealing, a random move can be swapping two cells, moving one cell, or swapping two I/O ports. The moving distance and swapping range of the cells are limited to a small number because the partitioning has already given good rough positions of the cells. After a move, procedure All_Cells() is called and area and wire length are evaluated to determine if the move is accepted. In fact, speed-up is possible, because a move may only affect part of the whole layout. For instance, when a cell is moved from X-position X_1 to X_2, only the part on the right side of Min(X_1,X_2) needs to be re-built.

3.2.3.3.6. The determination of the row number

In the previous discussion, how the algorithm works has been described, but a question is how the number of rows is determined. Assume a fixed outline of the SC module, S_X by S_Y, is given. The allowed height of all cell rows is:

$$H_{R-ALL} = S_Y - 2W_{PG},$$

in which, W_{PG} is the width of the power/ground rings. Recall that gaps may exist between rows to accommodate overflowing M3 spines. Fortunately the gaps are usually a very small portion of the row heights. Thus a simple way is to leave a small but fixed margin, 10% for instance, and get the row number:

$$N_R = \frac{H_{R-ALL}}{(1+0.1)\cdot H_C}.$$

The partitioning and All_Cells() are run to see if N_R is really a good choice and this iterates until N_R settles. Experiments have shown that the N_R computed with 10% row height margin works very well and almost no iteration of N_R adjustment is necessary. The width of the cell rows is a result of the run of All_Cells(). For fixed-outline designs, in fact no placement and routing tool, including ISCPD, can guarantee the success of creating a valid layout within the given outline. When the partitioning and one run of All_Cells() are completed, the width of the layout is examined. If it has already met S_X, more weight can be put on wire length in the following simulated-annealing improvement. On the contrary, if the width after partitioning is larger than S_X, the given module width, more weight can be put on area in the annealing. This forms a trade-off between the area and wire length.

If the outline of the module is not given, ISCPD uses

$$N_R = \frac{\sqrt{\frac{totalCellArea}{u}}}{(1+0.1)H_C}$$

to determine the row number. Parameter u is the estimated area utilization, and it can be a constant or derived based on the discussion in 3.2.3.4.1. In fact ISCPD can always produce a valid layout no matter what u is used, but a carefully chosen u prevents creating layouts with awkward shapes.

3.2.3.4. Practical issues

3.2.3.4.1. Large circuits

Virtually ISCPD does not have any limit on circuit size. However three-layer interconnection may become a limit as circuit size grows. Rent's rule gives an estimate on the relationship between the number of the connections and the number of gates in a region [LAND71]. A rough estimation is made as follows. The center region of a layout containing half of the total cells is examined. Rent's rule predicts that there are

$$N_{EC} = r_1\left(\frac{N_C}{2}\right)^{r_2} = 1.85 N_C^{\ 0.7}$$

external connections where r_1=3 and r_2=0.7 are constants. The region can provide maximally

$$N'_{EC} = 2\sqrt{\frac{(N_C/2)A_C}{u}}\left(\frac{1}{Pitch(2)} + \frac{1}{Pitch(3)}\right)$$

connections to the external world on its periphery, in which A_C is the average cell area and u is the area utilization. With $N_{EC}' \geq N_{EC}$, the relationship between N_C and u is obtained:

$$u \leq (N_C)^{-0.4}\left[0.14 A_C\left(\frac{1}{Pitch(2)} + \frac{1}{Pitch(3)}\right)^2\right]$$

With the technology data in our experiment, u=0.75 when N_C=10000, and the available external connections are sufficient when N_C<4900. Obviously as the circuit size becomes larger, the area utilization must drop, or more white space should be introduced to get sufficient routing resources. Low utilization can be prevented by using more metal layers. Modern chips can have six or more metal layers. Our algorithm works with three metal layers; therefore large circuits must be partitioned first into smaller modules, 10k-cell modules for instance. After each module is placed and routed by ISCPD, modules are connected with layers above M3. The module level placement and routing can adopt a scheme called "Fishbone" [MO03a], which also makes use of spine net topology for better predictability and routability. ISCPD is very suitable for chip design flows that require the generation of standard-cell modules with small to medium sizes. The Module-Based design flow we propose is such a flow.

3.2.3.4.2. Power/ground

The power/ground rings have been shown in Figure 3.11. Usually for designs with long rows, special cells called "power cells" are inserted into all the rows and these are vertically aligned, forming a column, called "power stripe". A power stripe is simply one power lines and one ground lines that vertically cross all rows on M2. Its X location is planned by some power routing tool. Power stripes can be treated as the ribs of the bottom/top I/O ports and scheduled for routing in a similar way.

3.2.3.4.3. Clock nets

ISCPD cannot route clock nets using a spine topology while meeting special requirements like clock skew. But suppose a clock planner is available. It uses the sequences of the cells in the rows to plan the topologies of the clock nets, possibly inserting buffers in the rows wherever necessary [JACK90]. Then the vertical segments of the clock net are routed as if they are the ribs of the bottom/top I/O ports. A virtual row called "clock rib row" is created, storing all the vertical wire segments of the clock net and their expected X-positions. Let $X_{MAX}(clock\_rib\_row)$ always be the next wire in the row to be put in the layout. The horizontal wires of the clock net join the spine packing in their rows.

3.2.3.5. Experimental results

In the experiments, fourteen examples were tested. The examples are from the LSynth91 synthesis benchmark set; the cell netlists were generated by *SIS* through technology independent transformation and technology mapping [SENT92, BRAY90, LGSY91]. All clock nets were removed. The characteristics of the example are given in Table 3.1. A 0.35μm technology was used. The three metal layers, M1 to M3, have wiring pitches of 1.0, 1.2 and 1.5μm respectively. M1 has no preferred routing direction. M2 prefers vertical and M3 prefers horizontal. The widths of the power/ground rings are 10 μm, and the I/O height is 5.0μm. We compare our algorithm with Cadence Silicon Ensemble version 5.3 (SE) running on the same machine. The standard-cell physical design in SE includes row generation, I/O placement, cell placement (Qplace) and routing (Wroute). The user has to define either the module size or the utilization (ratio of the total cell area and the total row area) to enable the generation of the layout outline and the rows. All programs were run on a Sun Blade 1000 workstation. The ISCPD program was written in Java.

For each example, we tried to find the smallest but violation-free implementation using SE. The row utilization was gradually reduced until the layout was fully routable. The minimum resolution of the row utilization for this search was 5%. ISCPD was run only once for each example. If in an example the ISCPD area was at least 5% larger than the SE area, we ran SE again using the ISCPD area just to see how the wire length may change. These are labeled SE* in the data tables. Since we tried to find the smallest area without routing violation for SE, we put 90% weight on area and 10% on wire length in the ISCPD annealing.

Table 3.1: Testing examples

circuit	#gate	#eqi.gate	#I/O	#net	#pin
mm4a	181	277	11	188	493
C3540	1342	1641	72	1392	3700
s5378	1767	3030	81	1802	4700
des	3558	4522	501	3814	10606
dsip	3647	6172	425	3875	10242
C5315	4908	6307	301	5086	14294
C7552	5932	7162	313	5825	16121
i10	6778	8238	481	6554	18169
C7552L	8461	10793	313	8667	24464
C6288S	8493	10091	64	8525	23294
C6288	11921	14399	64	11889	33188
mult32	12419	13238	96	12451	32266
S38417	16633	29786	134	16661	43937
clma	24497	45455	464	24879	80242

Table 3.2: Area and total wire length

circuit	area ($10^3\mu m^2$) <note>				total wire length (mm)			
	ISCPD part.	ISCPD SA	SE	SE*	ISCPD part.	ISCPD SA	SE	SE*
mm4a	24.4	23.3	24.9		6.4	6.1	7.1	
C3540	209	200	178	191	88.4	74.8	92.8	84.1
s5378	400	360	327	362	127	149	124	130
des	1115	1095	937	1088	677	612	629	610
dsip	827	810	778	810	659	625	587	594
C5315	1539	1467	1953		812	683	822	
C7552	1866	1795	1777		898	899	753	
i10	2359	2259	2039	2214	1213	1144	979	1103
C7552L	1902	1874	2081		733	751	565	
C6288S	3941	3727	3681		1716	1721	1650	
C6288	3496	3412	3538		1339	1255	882	
mult32	2453	2324	4059		1124	972	1023	
s38417	6869	6764	8512		3192	2785	2795	
clma	24298	23391	24794		11193	9815	10489	
compare	0.99	0.96	**1**		1.12	1.05	**1**	1.01

<note> the power/ground rings and I/O height may noticeably affect the areas of the small circuits.

Table 3.3: Wire segments and vias

Circuit	#segment (10^3)				#via (10^3)			
	ISCPD part.	ISCPD SA	SE	SE*	ISCPD part.	ISCPD SA	SE	SE*
mm4a	0.9	0.9	1.1		0.9	0.9	0.8	
C3540	7.4	7.4	8.8	8.5	7.4	7.4	7.0	7.0
s5378	9.8	9.7	10.5	10.5	9.4	9.4	8.4	9.4
Des	22.6	22.4	29.7	28.6	21.4	21.5	23.3	23.3
Dsip	22.3	22.3	28.0	27.5	20.4	20.5	21.4	21.5
C5315	30.2	29.7	36.9		28.7	28.6	29.2	
C7552	34.0	33.8	40.6		32.4	32.4	31.6	
i10	38.8	38.6	47.3	47.3	36.5	36.6	36.7	36.8
C7552L	46.9	46.7	51.9		46.6	46.6	40.9	
C6288S	51.5	51.4	63.4		49.0	49.1	49.6	
C6288	68.3	68.0	75.5		66.4	66.4	60.7	
mult32	66.0	66.0	75.8		66.5	66.5	62.7	
s38417	95.0	95.3	103		88.0	87.9	85.9	
Clma	193	192	204		160	160	172	
Compare	0.85	0.85	**1**	0.98	0.99	0.99	**1**	0.96

Table 3.4: Run time (minutes)

circuit	ISCPD part.	ISCPD SA	SE place	SE route	SE* place	SE* route
mm4a	0	0.01	0.07	0.64	0.71	
C3540	0.05	0.63	0.16	0.68	0.84	0.16
s5378	0.08	1.08	0.2	0.72	0.92	0.22
des	0.75	6.7	0.49	1.15	1.64	0.52
dsip	0.43	6.13	0.49	0.75	1.24	0.5
C5315	1.03	7.5	0.7	0.7	1.4	
C7552	2.23	9.2	0.93	1.23	2.16	
i10	1.7	7.7	1.18	1.27	2.45	1.13
C7552L	2.5	10.3	1.33	1.38	2.71	
C6288S	5.19	13.6	1.72	1.47	3.19	
C6288	5.73	15.6	2.1	1.79	3.89	
mult32	5.53	16.5	2.35	2.64	4.99	
s38417	11.2	23.3	18.7	97.6	116	
clma	56.7	44.4	25.8	234	259	
compare	0.63	2.82	1		0.89	

The areas and total wire lengths are reported in Table 3.2. In Table 3.3, the number of wire segments and the number of vias are reported. These data affect manufacturability issues [LAVI02, PLUM00], although we do not have any quantitative measurement of this. Run times are given in Table 3.4. Only the time spent on the final run of SE is reported. The results of the initial layout generated through partitioning are reported in the "ISCPD part." columns; and the final results after simulated-annealing improvement are listed in the "ISCPD SA" columns.

The results show that on average the partitioning-based and simulated-annealing-improved ISCPDs are 1% and 4% respectively smaller in area than SE. In terms of total wire length, ISCPD-part and ISCPD-SA are on average 12% and 5% worse than SE. Comparing the results of ISCPD-part and ISCPD-SA, we find that the initial layout generated right after partitioning is already close to the final layout produced through simulated-annealing. So if only estimation is needed or run time is critical, one may simply run ISCPD-part only.

ISCPD-part and ISCPD-SA were better than SE in the comparison of the number of wire segments and vias by 15% and 1% respectively. In ISCPD, the number of vias is about twice the number of pins. An input pin needs one via getting up to the M2 rib and another via getting up to the M3 spine. An output pin goes up to the M3 spine through two vias. Therefore, the numbers of vias in ISCPD are quite close to twice the pin number. Small deviations may occur due to the connections to the I/O ports. A pin is associated with either a rib or a spine, consuming one segment. A pin also needs zero to two M1 segments to reach its rib or spine, consuming another zero to two segments. The results show that the number of segments in ISCPD is also close to twice of the pin number. The SE router uses a different net topology, most probably Steiner tree; so the numbers of segments and vias is not easily predicted. We expected to see that its segment number is close to the via number, but this did not occur. Careful study of the layouts shows that the SE router actually makes use of a lot more M1 resources than ISCPD to build the interconnections. Since M1 is bi-directional (although horizontal is preferred), making turns on M1 does not involve vias. This is the reason why the segments in SE are about 14% more than vias.

The run time of ISCPD-part is on average 37% shorter than that of a complete placement and routing of SE (note: ISCPD is in Java). The run time of ISCPD-SA can be longer. In the experiment, the ISCPD-SA takes 182% more time. However the user has the freedom in setting a time-out for the simulated-annealing run, and whenever the annealing is stopped, the layout is valid and only about -3% worse area compared to the final one can be expected.

The SE* columns give the results that were generated for examples in which ISCPD-SA produced at least 2% larger area than SE. In these cases, we observe how the increased layout area (on average 9% increase) may affect wire length, number of segments and vias and run-times in SE. The total wire length increases by 1% on average, while the number of segments and vias both decrease. The via numbers of SE* are even smaller than for ISCPD, which indicates that more M1 resources were used in routing instead of going up to M2 and M3. This is consistent with the fact that more area is allowed. However increasing area implies that the distances between the cells may increase as well, which results in longer interconnections. Finally the run time of SE* is 11% shorter than SE, probably because larger area makes it easier for both placement and routing in SE. A layout example is shown in Figure 3.13.

(a) ISCPD

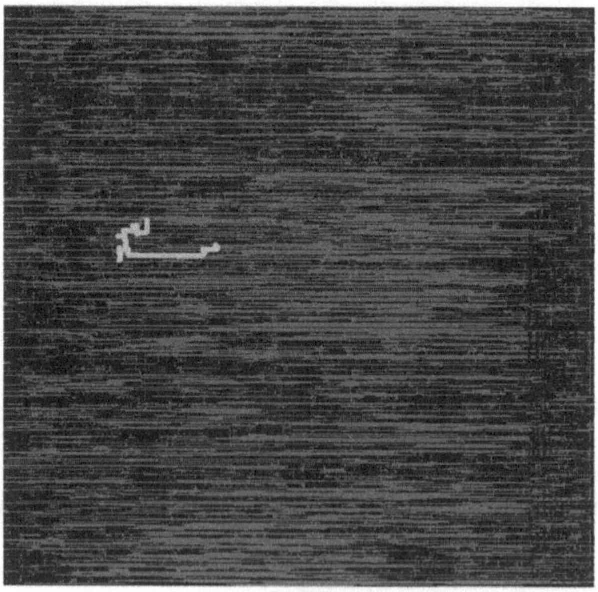

(b) SE

Figure 3.13: Layout pictures of C7552. A five pin net is highlighted.

3.2.3.6. Conclusion

The core of ISCPD is a procedure that translates the sequences of cells in the rows into a placement of the cells and the interconnections. The initial sequences are produced through a recursive bi-partitioning based on the spine net model. The layout can be further improved via simulated-annealing, although experiments show that the quality of the initial layout is already within about 3% of the final annealing result. This allows ISCPD to be used in different design stages which have different requirements for quality and run time. Full routability is always guaranteed. The algorithm needs no user input of the layout area or row utilization. Instead the area is an output of the algorithm, preventing iterations to search for the smallest area that is fully routable. The advantages of ISCPD include:

(1) ISCPD is an integrated algorithm without routability prediction used by most standard-alone placers. Routability is guaranteed by ISCPD.
(2) Run time can be traded for quality of the results, so user can get either an acceptable layout quickly or wait longer to get a better result. This allows flexible use of the algorithm in the design flow.
(3) The regular spine topology greatly simplifies the routing patterns. The numbers of wire segments and vias are both reduced. Being more regular, better manufacturing quality can be expected also.

3.2.4. Summary of the standard-cell structure

In this section, the most widely used standard-cell structure is described. The standard-cell structure is regular in the sense that the cells are placed in rows with the same height. The building blocks of the standard-cell design are the library cells, which are small pre-designed and pre-characterized modules implementing simple logic functions. Technology mapping is required in the logic synthesis to translate the Boolean network to a cell netlist. The physical design takes advantage of the fine granularity of the standard-cells, making possible effective and efficient placement algorithms.

After reviewing the standard-cell synthesis and physical design in the traditional design flow, an integrated standard-cell placement and routing algorithm called ISCPD was presented. The new algorithm improves the standard-cell physical design by removing inaccurate wiring estimation that may potentially cause routability problems. The option of adjusting quality versus run time offers flexibility to the design flow, in which different stages of the flow may have different requirements for quality and run time. The

ISCPD algorithm is adopted in the Module-Based design flow in Chapter 5 to construct the layout of the standard-cell modules.

3.3. PLA

In this section, the structure of Programmable Logic Array (PLA) is described. Although the direct use of PLAs in circuit design is limited to a few applications like simple instruction decoder, it is the basic element of more complicated circuit structures [MO03b], to be introduced in the next few sections.

The building block of a PLA is the NOR programmable array as illustrated in Figure 3.14. In Figure 3.14(a), a static configuration is depicted. The input signals, coming from the bottom, control the vertical lines in the array. The input buffers produce both the inverted and the non-inverted input signals. Each dot in the array is either a connected transistor controlled by the vertical line or a void transistor with no logical effect. The horizontal lines in the array create the wired-NOR function of the controlling signals. This is an application of a static NMOS gate, as described in Section 2.2.2.2. The left-side pull-up resistors are formed by depletion-type NMOS transistors. Despite its simple structure, the static NOR array dissipates static power. A dynamic configuration is depicted in Figure 3.14(b), which has been detailed in Section 2.2.2.3.

In addition to the signal inputs, there is a "start" signal determining whether the NOR is evaluating or pre-charging. The working periods of the input buffers and the pre-charging switches are non-overlapping, controlled by the "start" signal. In addition to the normal horizontal lines for outputs, a delay-tracking line produces a "done" signal (also low effective) indicating whether the evaluation is done. Therefore, dynamic NOR arrays can be cascaded; the "done" of one array driving the "start" of the succeeding array. The dynamic NOR array has the same functionality as its static counterpart, except that it needs an extra "start" signal to control the operation. But the dynamic version does not consume static power. Since static power is disadvantageous in modern IC design, we focus on dynamic NOR arrays.

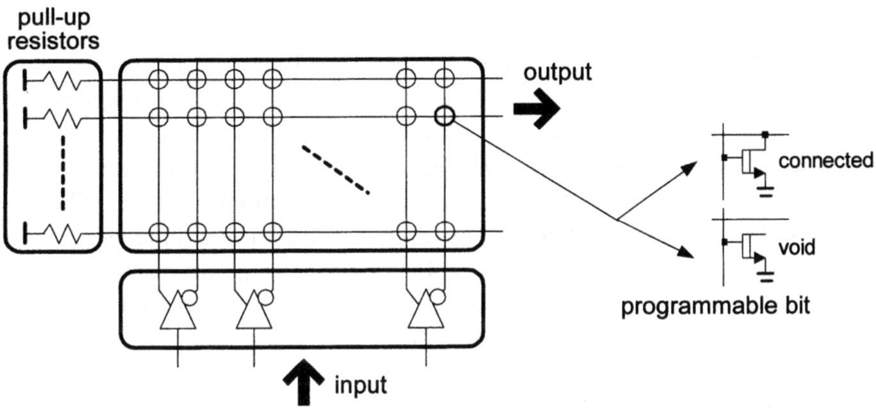

(a) static NOR array

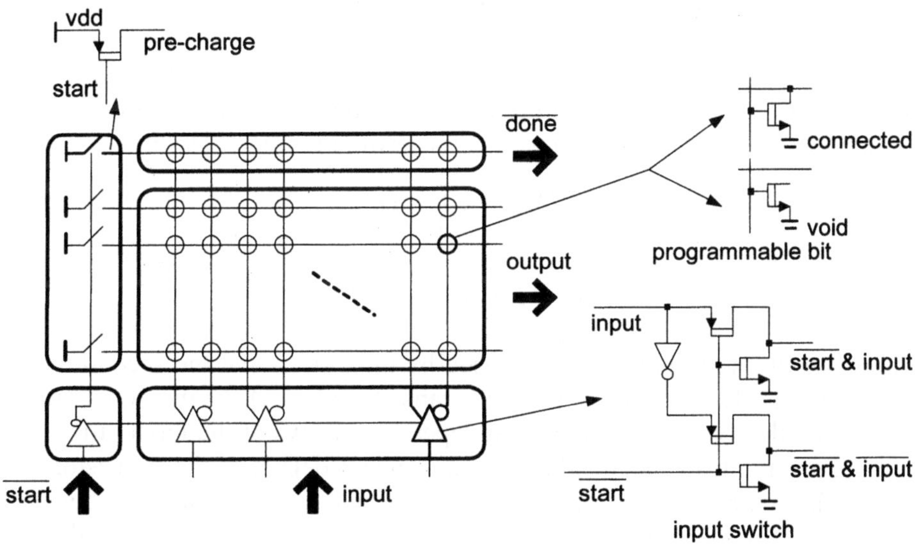

(b) dynamic NOR array

Figure 3.14: The NOR array.

The area of a dynamic NOR array, represented in terms of the number of inputs n_I and the number of outputs n_O, is:

$$A(n_I, n_O) = (W_{PC} + 2 \cdot W_V n_I) \cdot [W_{IB} + W_H \cdot (n_O + 1)].$$

In the formula, W_{PC}, W_{IB}, W_V and W_H are the widths of the pre-charging circuits, the input buffers, a vertical line and a horizontal line, respectively. The "2" in the equation reflects that an input variable is binate. For unate inputs, this might be dropped; but this may require some modification of the layout. The "+1" term in the equation is for the delay-tracking line. When the logic embedded in the array is given, the area is known immediately.

Since the input buffers are enabled by "start", the input signals begin to control the transistors at the same time. Therefore, the delay of the array is determined by the largest arrival time of the output signals:

$$D = \max_h \{D(h)\}.$$

$D(h)$ is the arrival time at output h given by:

$$D(h) = \max_v \{D(v,h) | Tr(v,h) = 1\},$$

in which, $D(v,h)$ is the delay from input v to output h. $Tr(v,h)$=1 indicates a transistor connects vertical line v to horizontal line h; otherwise $Tr(v,h)$=0. $D(v,h)$ is assumed to be the worst case delay, i.e. only the connected transistor at that cross point turns on while all other connected transistors driving the same horizontal line h are off. Using a linear timing model, $D(v,h)$ is:

$$D(v,h) = \left(D_{IB} + d_{IB} \sum_{hh} Tr(v,hh) \right) + \left[D_{TR} + d_{TR} \left(L + \sum_{vv} Tr(vv,h) \right) \right],$$

where the first term is the delay caused by the buffer, and the second term is the "turn-on" delay of the connected transistor at (v,h). In the equation, D_{IB} and d_{IB} are the load independent and dependent delays of the input buffer, D_{TR} and d_{TR} are the load independent and dependent delays of the transistor. L is the unified external load attached to the output, which is usually the input capacitance of the buffer in the next array and the capacitance of the pre-charging circuit. Thus, the delay of a NOR array is totally determined by its embedded logic, that is, the $Tr(v,h)$'s.

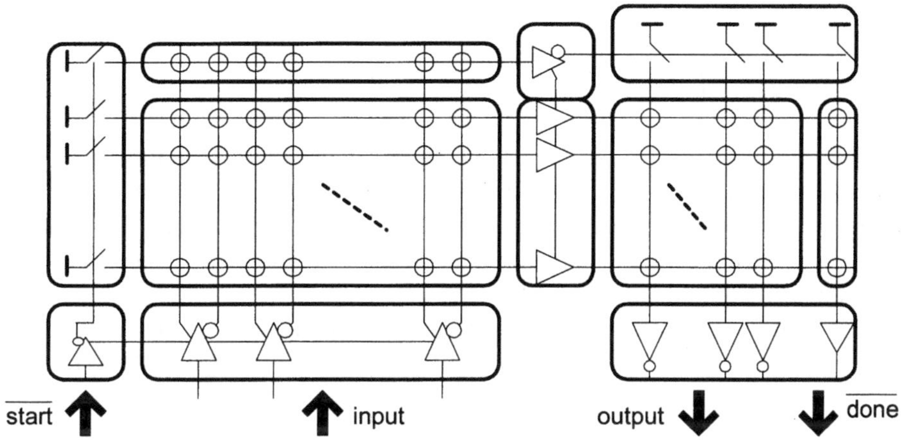

(a) The structure of PLA

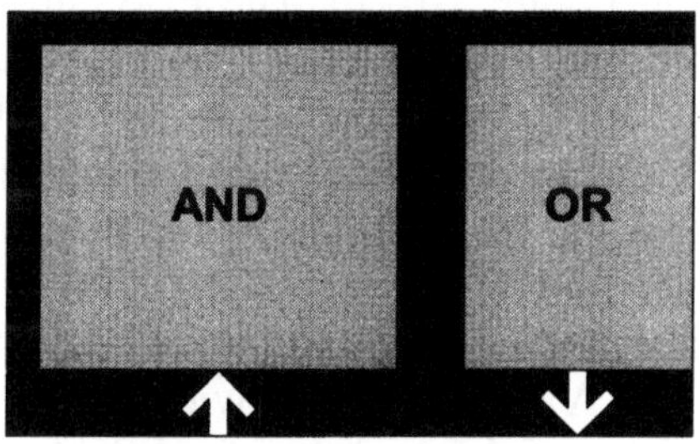

(b) abstract view

Figure 3.15: The PLA.

To implement a Sum-Of-Products (SOP), a PLA is built, as illustrated in Figure 3.15. The first half of the PLA, a NOR array (in fact NOT-NOR), is called the AND-plane and its output signals are called products. The second half (in fact NOR-NOT), is the OR-plane, and its output signals are called sums. Synthesis of PLAs has been well studied, and efficient algorithms exist [RUDE87, BRAY84]. The area and delay of a dynamic PLA can be easily derived from the equations for the NOR arrays. Thus, the delay and area of a PLA are fully determined by the embedded logic functions.

3.4. Network of PLAs

The Network of PLAs (NPLA) is a PLA-based structure that can implement larger circuit [KHAT00]. The basic idea is to cluster SOPs of a Boolean network into a set of multi-output PLAs. The design with NPLA involves a technology independent logic transformation, a clustering algorithm and a physical design stage.

The logic view of an NPLA is illustrated in Figure 3.16(a). Each black ellipse represents a cluster of single-output PLAs. A single-output PLA, a grey circle in the figure, is a SOP. The PLAs in the same cluster may share inputs, as shown in the zoomed picture on the top left corner of Figure 3.16(a). A cluster corresponds to a normal multiple-output PLA. To physically implement an NPLA, we have two approaches. One is the Field Programmable Logic Array (FPGA) style placement proposed by Khatri et al [KHAT00], as shown in Figure 3.16(b). The PLAs are placed in an array of slots. One slot can hold one PLA at most. The PLAs are connected through channel routing. In Figure 3.16(c), the macro-cell style implementation of NPLA is illustrated. A macro-cell placer using simulated-annealing or force-directed algorithm places the PLAs [MO00], and then an area router is called to make the interconnections. The area of the NPLA, if using FPGA style placement, can be easily derived by counting the number of slots in the array. The macro-cell style NPLA can usually provide higher area utilization (total PLA areas divided by the module area) than the FPGA style NPLA.

NPLAs have simple logic representations, that is, clusters of single-output SOPs. Therefore, no technology mapping is necessary, simplifying the task of the logic synthesis. Its physical design requires block-level (or macro-cell) placement and routing, which are usually more difficult than cell based physical design. NPLA maintains layout regularity inside each multiple-output PLA.

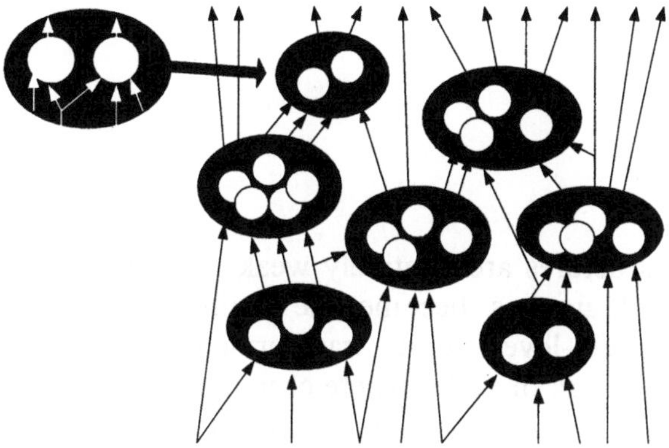

(a) the logic view

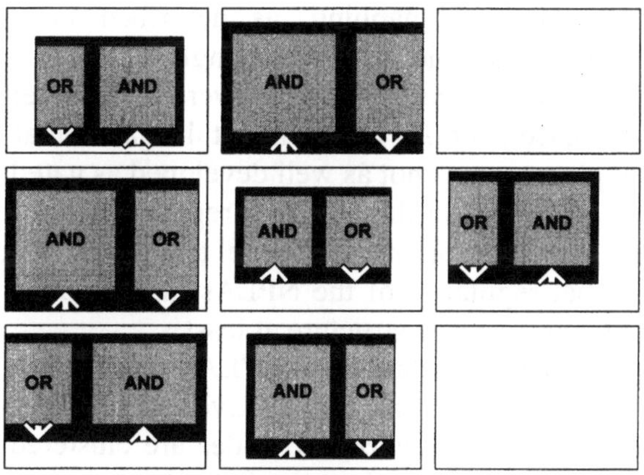

(b) NPLA (FPGA placement)

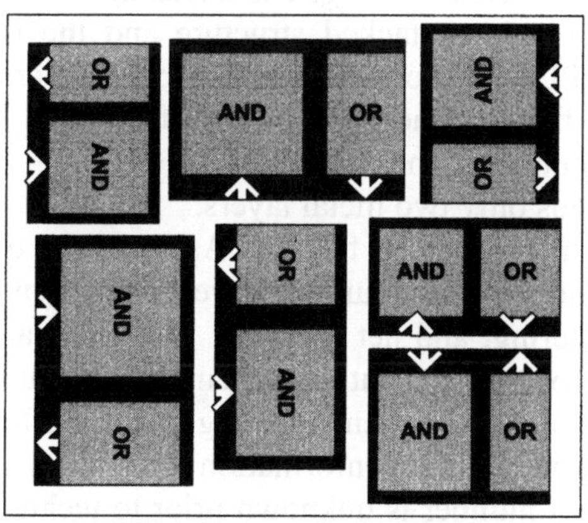

(c) NPLA (macro-cell placement)

Figure 3.16: The NPLA.

3.5. River PLA

3.5.1. Overview

Single-PLA structures are relatively weak in expressing logic functions. Thus multi-level logic has become the mainstream synthesis technique [BRAY90]. In multi-level logic transformation, the entire circuit is represented as a network of nodes where each node is a SOP logic function (a Boolean network). A common approach is to do logic independent transformation, and then map the circuit to a network of library cells via technology mapping. A natural step is to build an NPLA from the Boolean network without technology mapping, as described in the last section [KHAT00]. Thus some desirable features of single-PLAs such as technology-mapping free synthesis are preserved. However, NPLAs require placement and routing and the placement of PLAs is not at the gate but at the block level. So far, block level placement is not as well developed as gate level placement. Also, even though each PLA in the NPLA is regular, global regularity is low because of placement and routing (this can also require additional metal layers). In fact, global regularity of the NPLAs may be even worse than for standard-cells, which have a row structure at least.

To eliminate global irregularity of NPLAs caused by placement and routing, a multiple-PLA structure called River PLA (RPLA) is proposed [MO02b]. Given a Boolean network, the nodes are clustered into a stack of PLAs. In each PLA, the outputs and some of the input signals are ordered and fed to the next PLA via river routing. The advantages of RPLAs are:

(1) Since the RPLA is a stacked structure and the river routing is quite regular, the RPLA has both local and global regularity.
(2) The output buffer and the input buffer of the next PLA in the stack are merged, reducing area and improving speed.
(3) The RPLA needs only two metal layers.
(4) The design methodology for RPLAs consists of three main steps: technology independent multiple-level logic synthesis of a Boolean network, clustering, and net ordering. This maintains the advantages of the synthesis for NPLAs, but no placement and routing are necessary.
(5) At the technology independent stage, net buffering is complicated because there is no exact information about how a net will be routed; even the fanout number is unknown prior to technology mapping. Since RPLAs have dedicated buffers, which only drive a known number of loads, and river routing has fixed shapes and involves short distances,

the synthesizer is immune to buffering problems. The computation of the total delay becomes a simple summation of the delays of all the PLAs.

Thus, the RPLA is a good alternative to standard-cell. For future Deep Sub-Micron (DSM) designs in which regularity may be a key issue [BRYA01, MO02a], the local and global regularity of RPLAs should be an advantage.

A programmable version of RPLAs, called a Glacier PLA (GPLA), targets re-configurable applications, where a block of hardware is embedded in a System-On-a-Chip, and the functionality of the block is re-configured whenever the application changes. The name "Glacier" comes from the fact that the sizes of the PLAs and the river routing wires are "frozen". GPLAs, unlike Field Programmable Logic Arrays (FPGAs), have fixed wiring; thus they need no routing. Although FPGAs have a regular array structure, the topology of a routed net can be irregular, which is hard to predict during synthesis. The granularity of GPLAs is coarser than FPGAs, so to implement a chain of logic functions, the buffers involved in the GPLAs are smaller than in FPGAs. A simple example is an AND of 32 signals. Implemented with a GPLA, only one PLA is needed and only one pair of input and output buffers is involved. In contrast, to implement such a function with a 4-input Lookup-Table (LUT) FPGA, at least 3 pairs of input and output buffers are involved in the longest path. One may argue that the finer granularity of FPGAs leads to higher area utilization. Although true, a portion of the logic resources in FPGAs are wasted for either routability or performance considerations. In contrast, GPLAs waste area only because no appropriate logic can fill in. The design methodology for the FPGAs is similar to that for the standard-cells, except the technology mapping targets a LUT structure. Therefore, they also have problems with buffering and routing prediction.

The remainder of this section is organized as follows. The circuit structure is described in Section 3.5.2. The design methodologies for the RPLAs and the GPLAs are discussed in Section 3.5.3. Experimental results are presented and discussed in Section 3.5.4, and Section 3.5.5 gives a summary.

3.5.2. The structure of the RPLA

As shown in Figure 3.17, an RPLA consists of a stack of multiple-output PLAs. Since some of the nets feed more than one PLA in the stack, the output signals of a PLA may come from both the OR-plane (defined as OR-signals) and the AND-plane (defined as AND-through-signals). Those nets ending at the current PLA are called AND-only-signals. The AND-through-signals and the OR-signals of each PLA are ordered and connected to the next PLA in the stack via river routing. Metal layer one and two are assigned to the routing of the AND-though and OR signals respectively. The interconnections via river routing are local and hence short. The left sides of the PLAs are aligned. The primary inputs come in at the bottom. The primary outputs can be retrieved from the right side from any one of the PLAs or from the top. When they are derived at the right side, river routing (still on metal levels one and two) is used to bring the outputs of the PLA OR-planes to the right boundary.

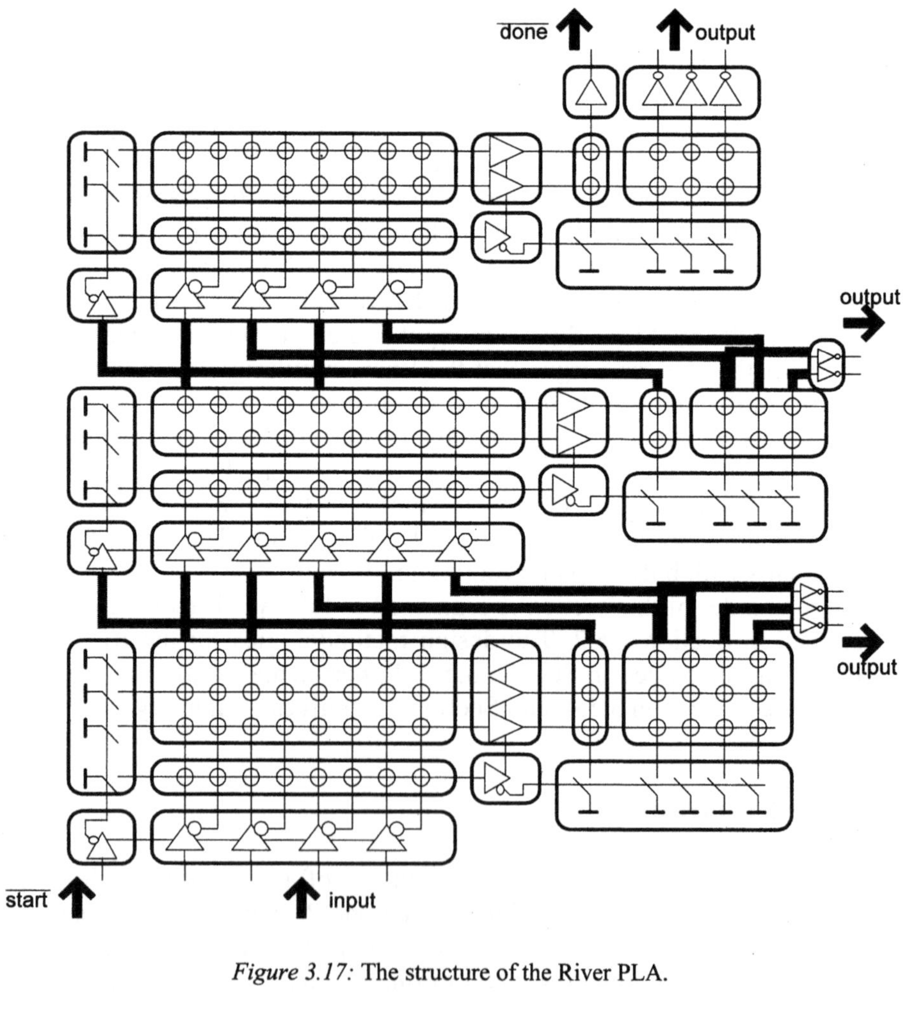

Figure 3.17: The structure of the River PLA.

A RPLA structure can be characterized by the following parameters:

n_L: the number of the levels.

$n_T(s,t)$: a $(n_L+1)\times(n_L+1)$ matrix, the number of the nets starting at level s and reaching level t, where $s \le t$. The primary input nets have starting level of 0. $n_T(t,t)$ denotes the number of outputs of the PLA on level t. It is conceivable that $n_T(s,t) \ge n_T(s,t')$ for $t \ge t'$, because with increasing level, part of the nets coming from lower levels are no longer used by any nodes, while the nodes on the higher levels create their own outputs, which become the new nets.

$n_I(t)$: the number of inputs to the PLA on level t., which is equal to $\sum_{s=0}^{t-1} n_T(s,t)$.

$n_O(t)$: the number of outputs of the PLA on level t, which is exactly $n_T(t,t)$.

$n_P(t)$: the number of product terms on level t.

For the GPLA, limits exist for each of the parameters. We use capital letters to define the limits: N_L is the limit for the number of levels, $N_T(s,t)$ is the limit for the number of the nets starting at level s and reaching level t, and $N_P(t)$ is the limit for the number of product terms on level t. It makes no sense to have a GPLA composed of PLAs with different width. Because of physical design issues, a column in the AND-plane (actually containing two bits, one for the positive signal, one for the complement) and a bit in the OR-plane may have different widths. To achieve a uniform PLA width, it is required that all

$$N_T(t,t) + r_{OA} \cdot \sum_{s=0}^{t-1} N_T(s,t) \qquad t = 1,2,.., N_L$$

are nearly the same, where r_{OA} is the width ratio of an OR-plane column to an AND-plane column. Normally this ratio is about 0.5. It is convenient to express the net and product term distribution by a consume/provide diagram as shown in Figure 3.18.

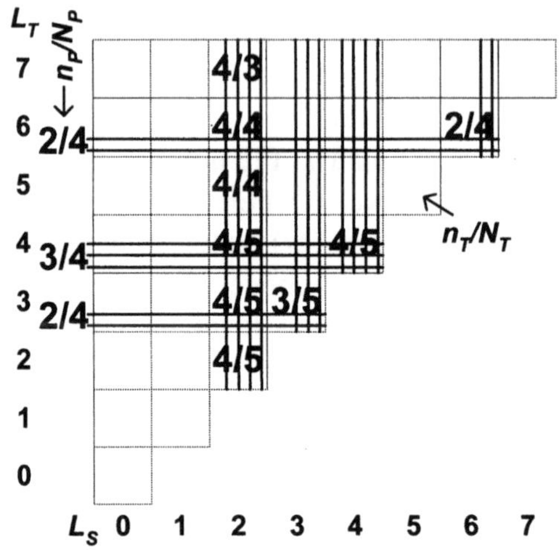

Figure 3.18: The consume/provide diagram.

This representation will be frequently used in the discussion of the GPLA synthesis algorithm. A valid GPLA implementation should satisfy $n_L \leq N_L$, $n_P \leq N_P$ and $n_T(s,t) \leq N_T(s,t)$ for all $1 \leq t \leq n_L$ and $0 \leq s < t$.

The area and delay of a RPLA, which can be easily derived based on the formulations in 3.3, are both explicitly expressed in terms of the PLA contents. Hence the objective function can be evaluated with good accuracy. This contrasts with standard-cell design where the area and delay are hard to predict at the technology independent stage. Although NPLAs also have better delay prediction than standard-cells, area estimation is hard, because it depends on the geometric characters of the PLAs and the quality of the block level placement tool.

The width, height and area of the PLA on level t are:

$$s_X(t) = W_{PA} + W_I \cdot n_I(t) + W_{BM} + W_O \cdot [n_O(t) + 1]$$
$$s_Y(t) = W_{BI} + W_P \cdot (n_P(t) + 1) + W_R \cdot [n_O(t) + 1] \quad .$$

The number of horizontal wires in the river routing region is exactly $n_O(t)$, as will be shown in Section 3.5.3. The "+1" terms in the formulae account for the delay-tracking lines. All the constants are defined in Table 3.5.

Table 3.5: Definitions of PLA parameters

W_{PA}	Width of pre-charging circuit of the AND plane
W_I	Width of a pair of complementary input columns
W_{BM}	Width of the intermediate buffer
W_O	Width of an OR plane column
W_{BI}	Width of the input switch
W_P	Width of a product line
W_R	Width of a wire in the river routing area

The height of a RPLA is the sum of the heights of the PLAs, and the width is the largest width:

$$S_X = \max_{t=1}^{n_L} s_X(t)$$
$$S_Y = \sum_{t=1}^{n_L} s_Y(t) \quad ,$$

The non-uniformity of the PLA widths results in white space on the right side of those PLAs whose widths are smaller than the RPLA width. The GPLA has the same area formulation, except everything in the formulae is constant.

The RPLA delay is just a summation of the PLA delays as formulated in Section 3.3, because of the dynamic configuration. The delays of the river-routing wires are negligible since they are local connections and have no vias.

3.5.3. The design methodology

The design flows of RPLAs and GPLAs share some common steps, because of their structural similarity. However, for RPLAs, the objective is to find a solution that reduces the total area, delay or combination of the two. For GPLAs, the first goal is to find a valid solution.

3.5.3.1. The synthesis of RPLAs

The design flow for RPLAs contains three steps: multi-level logic tranformation, node level-placement and net ordering. An objective function can be evaluated with good accuracy, because the area and delay are fully determined by the logic embedded in the RPLA. In contrast, the areas and delays for standard-cell designs are hard to predict during technology independent logic transformation.

3.5.3.1.1. Multi-level logic transformation

A multi-level logic synthesis algorithm is called to generate an initial Boolean network [SENT92].

3.5.3.1.2. Node level-placement

Node level-placement is possible because some nodes have flexibility in their levels. The single-output nodes in the network are levelized. Note that a node is an SOP (in general, two logic levels) structure, but the term "level" here means the level of the node in the Boolean network, unless otherwise mentioned; the real number of logic levels can be twice the number of network levels. The number of PLAs in the RPLA is equal to the number of network levels. The nodes on the same level are clustered into a multiple-output PLA in the RPLA. After clustering, each PLA is reduced further by *ESPRESSO*.

Suppose the level of node d is $l(d)$. If $l_{HFI}(d)+1<l(d)$ or $l(d)<l_{LFO}(d)-1$, where $l_{HFI}(d)$ is the highest fanin level and $l_{LFO}(d)$ is the lowest fanout level, then the node is said to be down-movable or up-movable. In this case, the level of a node d can be chosen within the range of $l_{HFI}(d)$ and $l_{LFO}(d)$. Notice that the flexibilities of different nodes may be correlated, because of the fanin/fanout relations. This is like the slack computed in delay analysis where each node has delay 1. It has the good effect of decreasing the number of possible combinations, and thus shrinking the solution space to be searched. As illustrated in Figure 3.19, nodes A, B and C are flexible. For example, A can have level 3 or 4. The flexibilities of different nodes can be dependent. For example, B can be put on level 1, 2 or 3, and C can be put on level 2, 3 or 4. However, C must always be on a higher level than B. This is actually not all bad news, because it decreases the number of possible combinations. In the figure, the total number of combinations is not 2×3×3=18, but 2×6=12.

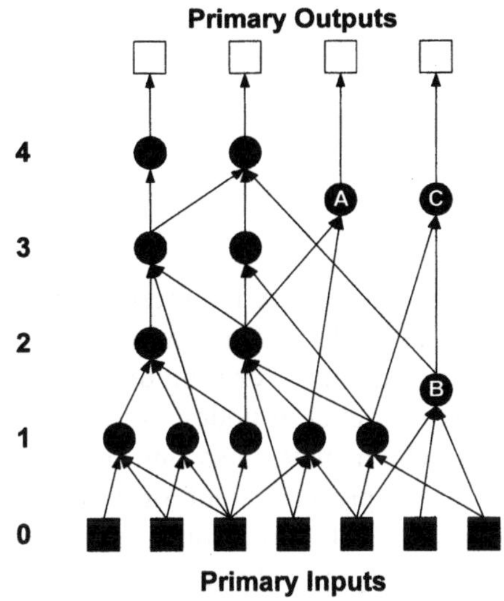

Figure 3.19: An example of the node level flexibilities.

Simulated annealing is suitable for the node level placement, since the solution space is reduced due to the level dependencies, and the evaluation of area and delay is straightforward. The objective function is chosen as a weighted sum of the area and delay. In the area and delay computation mentioned above, every variable can be trivially derived from the configuration, except n_P, the number of product terms. The exact value of n_P requires a SOP minimization [RUDE87]. One could call the SOP minimizer each time the cost is to be computed to get the exact product number; this might not be slow, because the PLA in a RPLA is usually not big. A faster but more conservative way is to use ns_P, the sum of the product numbers of all the nodes in the cluster, which is an upper bound of the number of product terms of the final multiple-output PLA. A trade-off is that when ns_P exceeds a threshold, we do not call the SOP minimizer but simply use the value ns_P. Otherwise a SOP minimization is performed.

Another issue is the area of the river routing region, which affects the total area of the RPLA. The net ordering algorithm, to be discussed later, will guarantee that the thickness of a river routing region, or the number of horizontal wiring tracks between two PLAs, is always linear with the number of outputs from the OR-plane of the previous PLA. The net ordering algorithm might increase the widths of some PLAs by a small amount. But as long as the PLAs affected are not the widest in the stack, this has no effect on the total area of the RPLA.

After the level of the flexible nodes is determined, the nodes on the same level are assigned to a multiple-output PLA and further processed with a SOP minimizer.

3.5.3.1.3. Net ordering

The sequence of nets in the RPLA is determined such that river routing between any pair of adjacent PLAs is feasible. Let $v_S(n)$ be the starting level of net n, which is the level of the node driving the net. Let $v_E(n)$ be the ending level of net n, which is highest level of nodes using n as a fan-in. In the RPLA, a net only has one wire, which may cross one or more AND-planes of the PLAs. If a wire starts from the primary input, or $v_S(n)=0$, it is simply a vertical segment reaching level $v_E(n)$. If it starts from the output of the PLA on level $v_S(n)>0$, it first turns into the region between PLAs $v_S(n)$ and $v_S(n)+1$, then go vertically from $v_S(n)+1$ through $v_E(n)$. Although some vertical wire segments in the AND-planes traverse more than one PLA, they are split by the buffers in the PLAs. Therefore all the connections are short. The net ordering algorithm first finds a packing of the vertical segments in the AND-planes of the PLAs. The ordering of output signals of the PLAs follow their ordering appearing in the AND-planes, and the output signals reach their vertical segments via river routing immediately after they leave the PLAs generating them. Hence the thickness of the river routing region between PLAs on level v and $v+1$ is determined by the number of outputs of the PLA on level v. A minor issue is that 1 is added to the thickness to account for the "done" signal.

Following is a brief explanation of the algorithm. First, order the nets in descending order of $v_E(n)$. For nets with the same $v_E(n)$, order them in ascending order of $v_S(n)$. There is no restriction on the nets with the same $v_S(n)$ and the same $v_E(n)$, since they always go side by side. Then from left to right, greedily fill in each vacant slot with the next available segment in the ordered list. The greedy algorithm is equivalent to the left-edge algorithm used in channel routing [SHER93], which gives an optimum packing of the vertical wire segments in the AND-planes of the PLAs. Finally the widths of the PLAs are derived. The width of a PLA is determined by the number of output signals, which is a constant, and the width of the AND-plane. Note that the packing algorithm may generate some empty space in the AND-plane, which causes the width of the PLA to be larger than the estimated value. As long as such PLAs are not the widest in the RPLA stack, all the computations remain the same. Experiments indicate that the case where the widest PLA has empty columns is rare.

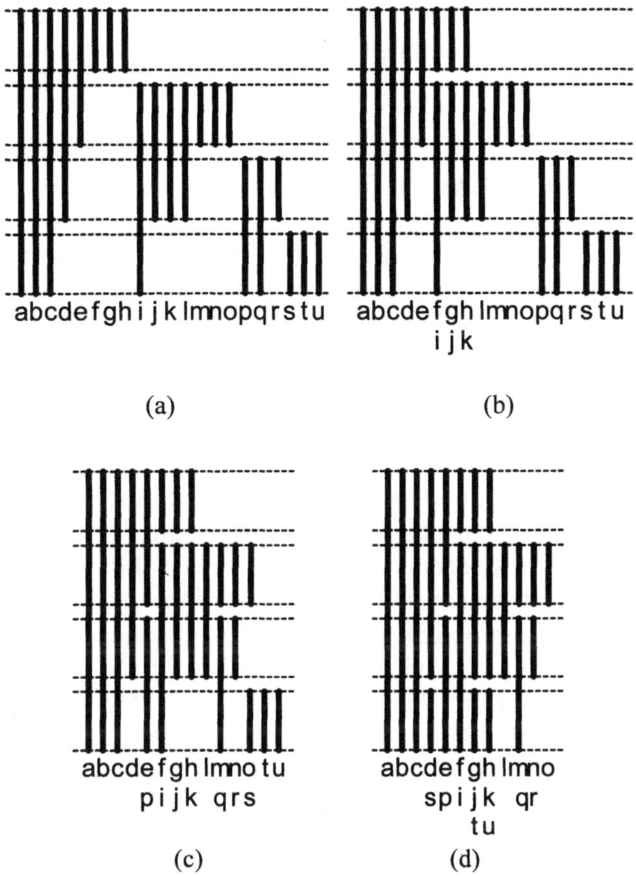

Figure 3.20: An net ordering example.

The net ordering algorithm can be explained via the example shown in Figure 3.20. First, order the nets in the descending order of $v_E(n)$. For the nets with the same $v_E(n)$, order them in ascending order of $v_S(n)$. This results in Figure 3.20(a). There is no restriction on the nets with the same $v_S(n)$ and the same $v_E(n)$, and they will always go side by side. Then, starting from left to right, put each net in the leftmost slot that is vacant. Figures 3.20(b) to (d) illustrate the process. Applying this algorithm will not generate any twisting in the routing layers, meaning that river routing with two metal layers is feasible. This simple scheme causes a little inaccuracy in the computation of PLA widths. Notice that in the bottom PLA in Figure 3.20(d), there is a vacant slot at the position "l". This makes the real AND column number of the bottom PLA larger than the number of its inputs, which is used to calculate the PLA area as described in Section 3.5.2. Since the width of the bottom PLA is not the largest, this estimation error does not cause any error in computing the RPLA area.

3.5.3.2. Synthesis of GPLAs

The synthesis of GPLAs starts with a multi-level logic transformation with level control. This step is similar to that in RPLA synthesis, except the number of levels is controlled to be no greater than N_L. It is followed by an iteration of node level-placement and methods to fix different violations. The iteration either completes with a valid solution found, or quits after a given number of trials and switches to a larger GPLA configuration. If successful, a post processing is performed to improve speed. Post processing tries to improve the performance, because the previous steps focus on seeking a valid rather than a fast implementation. Post processing is a greedy approach, accepting any node displacement that preserves validity but reduces total delay. The algorithm terminates when no further improvement can be found.

We focus on the problem of how to fix violations. In the following context, we use the term "N_P violation" to define the violation of $n_P > N_P$, "N_A violation" to define $n_T(s,t) > N_T(s,t)$ where $s < t$, and "N_O violation" to define $n_T(t,t) > N_T(t,t)$. As in the RPLA design flow, simulated annealing is used to place the nodes on levels. However, the cost function is the total number of violations. Subsections 3.5.3.2.1 through 3.5.3.2.4 describe the methods for fixing violations remaining after node level-placement.

3.5.3.2.1. Net shifting

Net shifting provides "detours" for nets causing N_A violations. Starting from the bottom left of the provide/consume diagram, we handle N_A violations one by one. The following pseudo code describes how a N_A violation of $e_V = (n_T(s_V,t_V) - N_T(s_V,t_V))$ excessive signals at (s_V,t_V) is handled by net shifting:

```
for (t=tV-1; t>sV, t--) {
    sn=min(NT(t,t)-nT(t,t), NP(t)-nP(t));        //no. of spare lines.
    if (sn≤0) continue; else dn=min(sn, eV); //no. of detours.
    nT(sV,tt)-=dn for all tt=t+1 to nL;
    nT(t,t)+=dn; nP(t)+=dn; nT(t,tt)+=dn for all tt=t+1 to nL;
    if (eV==0) return success;
}
return failure
```

An example is shown in Figure 3.21, where the N_A violations at (s=2,t=5) and (s=2, t=7) are eliminated through net shifting.

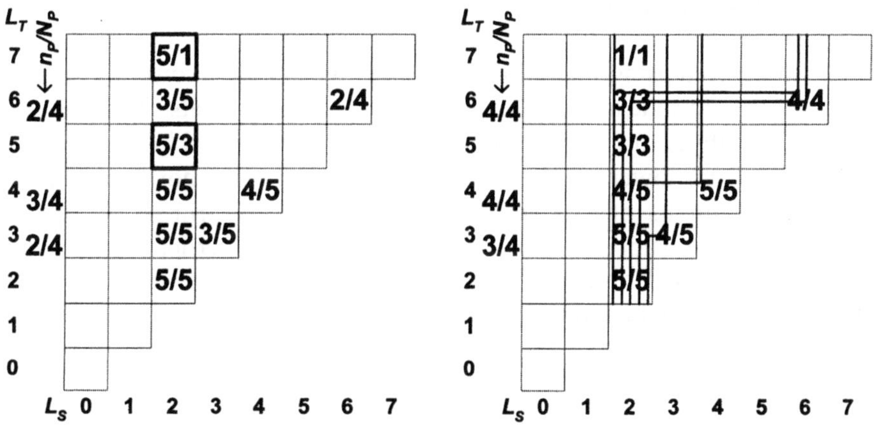

Figure 3.21: An example of net shifting.

3.5.3.2.2. Node decomposition

The objective of this operation is to reduce the number of product terms on a level with the N_P violation. The decomposition operation factors a node and makes its divisors new nodes in the network. For example, consider the node function F=e+fg+fh which consumes 3 product lines. If F is decomposed into F=e+fx and x=g+h, then F only consumes 2 product lines. So if there is a shortage in product lines on the level of F, the above decomposition saves one line, but at the expense of creating a new node. However the new node may not have trouble in finding a place in the stack. Area targeted multi-level logic synthesis aims at minimizing the total number of literals, which is a good estimate of the total number of transistors if the circuit is to be implemented with gates. In the above example, the decomposition does not change the total number of literals, so the initial synthesis step may not generate this decomposition. However, in the synthesis of GPLAs, the first goal is to get a valid implementation for the given configuration. Using more nodes or more literals should not be a problem as long as the mapping is valid. To choose the best node to be decomposed, we simply examine the nodes on the level with the highest N_P violations and select the one having the most product terms. Since decomposition of a node will generate new nodes and the new nodes may affect other levels, the node decomposition is only performed on one node at a time.

3.5.3.2.3. Node collapsing

The node collapsing operation deals with the N_A violations. N_A violations mean that the current and higher levels use excessive nets from the lower levels. A remedy is to decrease the need, of the current and higher levels, for nets coming from lower levels. The collapse operation removes intermediate nets from the inputs of a node and replaces them by primary inputs. This operation can be supported by reserving sufficient vertical lines for the primary input nets. However, the number of product terms of the collapsed node will increase, which may cause an N_P violation.

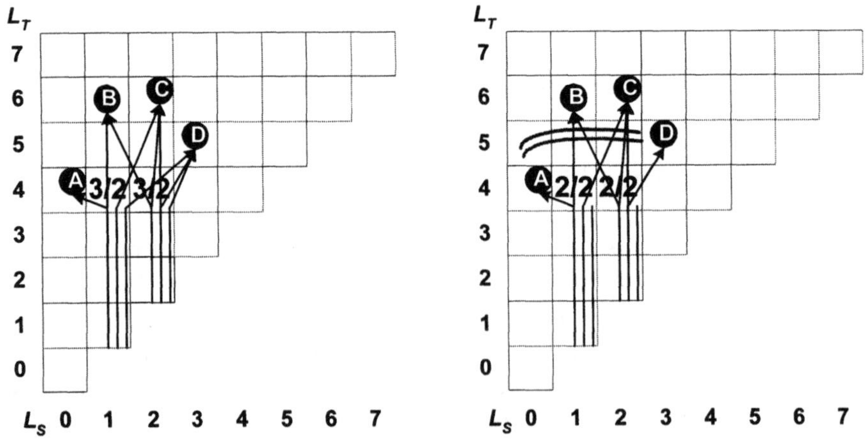

Figure 3.22: An example of node collapsing.

Simply collapsing a node using the most intermediate nets crossing the violation level is not a good idea, because other nodes on or above this level may use these nets. Rather we choose the node using the most unique intermediate nets crossing the violation level. The number of unique nets is the saving achieved by collapsing the node. The algorithm checks all nodes on or above the level with the highest N_A violations. A node with the maximum number of unique nets crossing the violation level is identified and collapsed. An example is shown in Figure 3.22, where node D is collapsed because it used two unique nets passing through level 4.

3.5.3.2.4. Node elimination

N_O violations are handled via node elimination. This operation eliminates a node and collapses its function into all its fanout nodes. Eliminating a node immediately reduces $n_T(t,t)$ by one. Node elimination effectively makes a copy of the current node function in all its fanout nodes, so the implementation of the functions in the fanout nodes may cause unexpected growth in the product terms. Thus the node to be eliminated should be chosen carefully. The first criterion is the one with the least fanouts. If a tie occurs, choose the one with the least product terms. Further ties can be resolved by choosing the one with the least input nets.

3.5.4. Experimental results

3.5.4.1. RPLA

In the experiments, three structures are compared: standard-cells with three-layer (3 metal layers) over-the-cell-routing (SC), four-layer NPLAs (NPLA) and two-layer RPLAs (RP). Each example starts with a logic synthesis with *SIS* on the initial Boolean network and creates a network with n_{L0} levels [SENT92]. Then the number of levels is gradually reduced via partial node collapsing. For each number of levels, n_L, the circuit is technology mapped twice with *SIS*, one with area priority and one with delay priority (*SIS* command "map -m 0 –AF" is used for area-prior tech-mapping, and "map -n 1 –AFG" for delay-prior tech-mapping), generating SC-a(n_L) and SC-d(n_L). We assume the SC designs have 100% area utilization and are 100% routable, so no placement and routing is performed. At the end of this chapter, a comparison will be made for all the structure discussed. The results of SC examples reported there involve a complete placement and routing. A clustering algorithm, similar to the one proposed in [KHAT00], is adopted to build the NPLAs; however there is no area or delay priority. The placement of the NPLAs is done by a force-directed macro-cell placer [MO00], producing NPLA(n_L). Two RPLAs are synthesized, one with 100% weight on area and one with 50% weight on area and 50% weight on delay, generating RP-a(n_L) and RP-d(n_L). A 0.35-micron technology is assumed for all circuits. The NPLA clustering and the RPLA mapping algorithms are written in JAVA; and all other procedures are within the *SIS* package and *ESPRESSO* [RUDE87]. The programs ran on a Dec Alpha 8400 5/625 workstation.

Sixteen combinational examples from the LGSynth 91 benchmark set [LGSY91] were tested. The gate counts of the examples range from 300 to 17K. The results are given in Table 3.6 and 3.7. To give a better view of the results, we normalize the area and delay data and plot them in Figure 3.23. For each example, the normalization is with reference to the data of the SC-a(n_{L0}). Results from the same example, but different area/delay tradeoffs, are connected with a single curve. The layouts of all three types of implementations for alu2(n_{L0}=10) are shown in Figure 3.24; routing is done by the *Cadence Warp Router*.

The experimental results show that in general, fewer levels mean faster speed but larger area. RPLAs (2 layers) needs on average 23% more area than SCs (3 layers), and NPLAs (4 layers) needs on average 31% more area than SC. Note that RPLAs and NPLAs do not use 100% of the area. RPLAs have an average area utilization of 78%, and 76% for NPLAs. If the module is to be embedded in a big chip, other circuits may occupy this white space. In terms of the delay, on average RPLAs are 4% slower than SCs, and NPLAs 9% slower. Figure 3.23 shows that SCs have more predictable shapes for the area/delay trade-off curves, while NPLAs and RPLAs have more scattered distributions. Logic synthesis times of SCs are several times faster than for NPLAs and RPLAs. However in the RPLA case, the synthesis time is exactly the final design time. For SCs and NPLAs, additional time is needed to complete the placement and routing. In addition, placement and routing may change the delay achieved by the synthesis; thus it is possible that the final layouts do not meet the requirements, which may invoke another run of the design flow. Circuit regularity is qualitatively shown in Figure 3.24, and intuitively it seems that RPLA gives the highest regularity among all the structures.

Table 3.6: Area comparison of SC-a, SC-d, NPLA, RPLA-a and RPLA-d

		cell count		Area (1000 um^2)					area utilization		
circuit	#level	SC-a	SC-d	SC-a	SC-d	NPL	RP-a	RP-d	NPL	RP-a	RP-d
alu2	10	644	875	34.8	47.2	15.0	14.1	14.4	80%	79%	76%
	6	888	1091	48.0	58.9	46.0	29.0	33.0	73%	77%	73%
	5	972	1222	52.5	66.0	55.0	38.4	38.4	82%	78%	78%
alu4	16	1175	1522	63.4	82.2	49.0	42.6	46.0	74%	75%	75%
	10	1494	1716	80.7	92.7	80.0	50.9	62.3	80%	78%	73%
	7	1958	2219	105	119	109	69.5	78.8	80%	83%	74%
	5	3119	3333	168	180	162	120	148	68%	80%	82%
apex6	7	1180	1572	63.7	84.9	93.0	85.9	86.3	92%	90%	87%
	4	1380	1575	74.5	85.0	99.2	122	119	88%	90%	89%
apex7	8	350	430	18.9	23.2	14.8	9.5	12.9	90%	97%	82%
	5	425	522	22.9	28.2	25.5	14.7	18.8	75%	98%	85%
C1355	10	755	900	40.8	48.6	70.8	61.8	61.8	72%	61%	61%
	7	813	947	43.9	51.1	79.4	50.1	50.1	77%	75%	75%
	5	1113	1477	60.1	79.8	89.3	77.2	80.2	70%	80%	77%
C2670	12	1119	1516	60.4	81.9	159	158	205	80%	75%	71%
	9	1733	2213	93.6	119	200	170	218	72%	79%	82%
	6	2150	2450	116	132	259	182	240	71%	69%	72%
C3540	23	2044	2780	110	150	150	136	140	81%	76%	75%
	13	3575	5286	193	285	203	189	200	73%	77%	73%
	9	4397	6363	237	343	315	200	234	71%	77%	70%
	6	6819	11091	368	599	504	310	389	68%	72%	76%
C5315	15	2633	3133	142	169	290	312	236	77%	73%	71%
	8	3500	4027	189	217	420	378	439	73%	91%	83%
	6	4444	5247	240	283	500	450	508	67%	98%	88%
C6288	25	6086	7861	328	424	694	725	732	80%	65%	65%
	20	7763	8813	419	476	1238	1083	1129	78%	73%	77%
	14	11883	16388	641	885	1785	1356	1759	79%	82%	79%
C7552	18	3388	5055	183	273	609	692	730	80%	61%	67%
	14	5283	6119	285	330	770	710	780	80%	77%	69%
	9	6805	9180	367	495	830	760	820	75%	72%	77%
	6	8352	10894	451	588	902	820	880	72%	80%	82%
dalu	14	1327	1691	71.7	91.3	130	127	164	79%	77%	72%
	8	1638	2000	88.5	108	174	174	175	77%	77%	78%
	5	2222	2472	120	133	207	140	181	65%	94%	84%
des	8	5038	5569	272	300	999	887	1093	82%	84%	82%
	5	11116	16511	600	891	1620	1292	1527	81%	78%	81%
i8	7	1633	2186	88.2	118	204	216	216	88%	79%	78%
	5	1713	1997	92.5	107	220	187	173	79%	86%	86%
i10	22	3611	4833	195	261	450	459	470	77%	71%	77%
	13	7777	9194	420	496	702	555	620	80%	80%	82%
	9	11388	13250	615	715	957	770	890	72%	82%	85%
	6	13916	17222	751	930	1103	902	1029	74%	92%	88%
k2	14	1638	2261	88.5	122	200	203	207	80%	76%	75%
	8	1913	2783	103	150	230	232	254	79%	86%	82%
	4	4050	5675	218	306	299	250	288	73%	90%	85%
x3	8	1413	1683	76.3	90.9	150	94.9	143	71%	89%	76%
	4	1522	1788	82.2	96.6	169	138	150	62%	90%	90%

Table 3.7: Delay comparison and synthesis time

		delay (ns)					Synthesis time (minutes)				
circuit	#level	SC-a	SC-d	NPL	RP-a	RP-d	SC-a	SC-d	NPL	RP-a	RP-d
alu2	10	8.0	6.1	8.5	8.5	8.3	0.5	0.5	1.9	2.4	2.6
	6	7.4	5.9	8.5	8.4	8.2	0.5	0.5	1.8	2.2	2.1
	5	7.2	5.6	8.1	8.2	8.0	0.5	0.5	1.6	2.2	2.2
alu4	16	15.6	15.0	16.5	16.6	16.3	0.5	0.5	2.3	3.1	3
	10	12.9	12.3	15	15.3	14.8	0.6	0.6	2.5	3	3
	7	12.0	11.2	14.3	14.2	13.3	0.8	0.8	3	2.9	2.8
	5	11.5	11	12.9	12.2	12.8	0.9	1	3	2.8	3
apex6	7	14	13.2	14	14.1	13.4	0.8	0.8	2.3	3	3
	4	13.6	12.1	12.3	12.3	12.1	1	1.1	2.4	3.4	3.5
apex7	8	6.4	5.8	6.6	6.7	6.4	0.4	0.4	1.8	2	1.9
	5	5.2	4.5	5.8	5.6	5.3	0.4	0.4	1.8	1.9	1.8
C1355	10	9.4	8.1	10.9	10.7	10.7	0.5	0.5	2	2	2.2
	7	8.8	7.9	10.0	9.4	9.4	0.6	0.5	2.2	2.8	2.5
	5	8.2	7.4	9.4	9.0	8.2	0.6	0.6	2.3	3	2.9
C2670	12	17.4	16.1	17.8	18.4	17.7	0.8	0.8	4	4.4	3.9
	9	17.0	15.9	17.0	18.0	16.4	1	1.1	4.2	4	3.9
	6	16.8	14.2	14.5	15.0	13.2	1.5	1.6	5	4.4	4.2
C3540	23	25.8	22.4	28.9	29.4	28.4	0.8	0.9	4.2	4.8	3.9
	13	20.0	19.9	22.4	26.4	20.8	1	1.2	3.3	4.9	4.2
	9	18.8	18.0	21.0	20.5	18.4	1.8	2	3.8	4.4	4.4
	6	17.3	16.4	19.4	17.4	15.4	2	2.4	4	4.4	4.3
C5315	15	29.0	27.0	30.8	31.4	30.5	1	1	4.9	4.8	4.2
	8	25.0	26.1	20.3	20.3	19.0	1.5	1.8	4.8	4.8	5.1
	6	22.0	25.0	22.0	22.7	19.6	2.2	2.4	5.6	5.3	5.9
C6288	25	50.0	44.0	55.8	55.1	54.6	5	5.3	8.4	9	10.4
	20	47.0	46.1	69.2	72.9	53.0	5	5.2	9	12.2	13
	14	45.2	42.2	60	62.2	53.1	5.8	5.8	10.2	12	10.5
C7552	18	50.4	40.2	40.2	43.9	30.5	3	3.1	8.2	8.2	9
	14	40.2	30.3	33.1	30.0	23.2	3.4	3.4	12	8	8.9
	9	30.0	25.4	25.4	25.0	19.0	4	4.1	11	9.5	9.9
	6	24.3	20.6	20.1	22.0	18.8	4.9	5	11.2	10	10.3
dalu	14	18.0	16.4	21.9	21.9	21.2	1.2	1.3	4	3.2	3.4
	8	16.9	16.0	21.3	21.2	18.2	1.4	1.4	4	3.3	3.9
	5	15.1	14.9	18.8	16.1	16.3	1.5	1.5	3.9	3.8	3.8
des	8	36.0	26.2	33.0	34.0	30.2	2	2.3	4.5	5.9	5.1
	5	26.4	26.0	21.4	20.0	14.0	4.9	4.8	6	8.3	8.9
i8	7	14.8	14.2	12.3	12.0	10.0	0.9	1	3.1	3.9	3.2
	5	14.1	13.7	11.2	10.8	8.9	0.9	1.1	2.9	3.3	4
i10	22	43.7	43.2	34.2	35	30.2	1.5	1.8	5.2	4.9	5.2
	13	35.6	32.5	28.2	28.2	22.3	2.9	3.3	5.9	5.1	5.5
	9	22.9	21.9	21.1	22.3	20.5	3.3	3.8	5.9	8.2	8.7
	6	19.9	14.3	22.0	22.4	14.8	5	5.5	6	9	9.1
k2	14	27.5	20.6	27.5	26.3	26.1	1	1.4	5.9	5	5.7
	8	20.2	15.0	20.1	21.4	18.2	2	2	6.4	6.8	6.4
	4	18.8	14.8	17.7	18	15.4	3.7	4.1	6	7.2	6.9
x3	8	14.8	14.2	14.4	15.8	14.2	0.7	0.8	1.8	2	2.1
	4	12.4	11.6	13.0	12.8	12.7	0.8	0.8	1.8	1.9	2

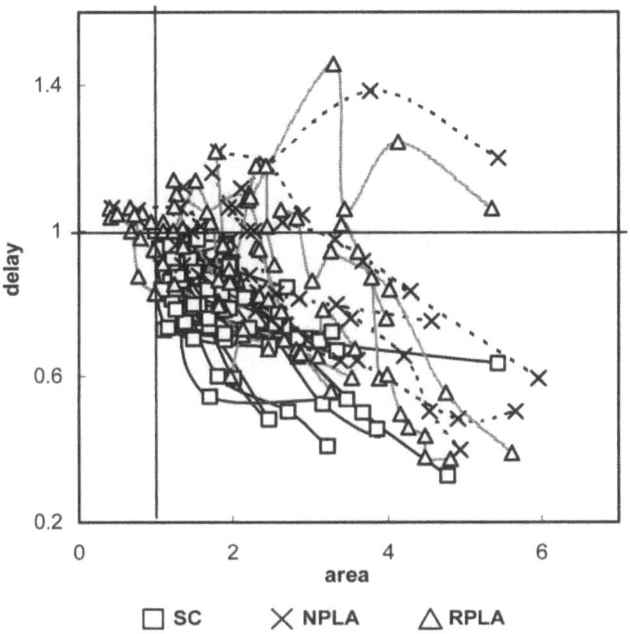

Figure 3.23: Normalized area/delay.

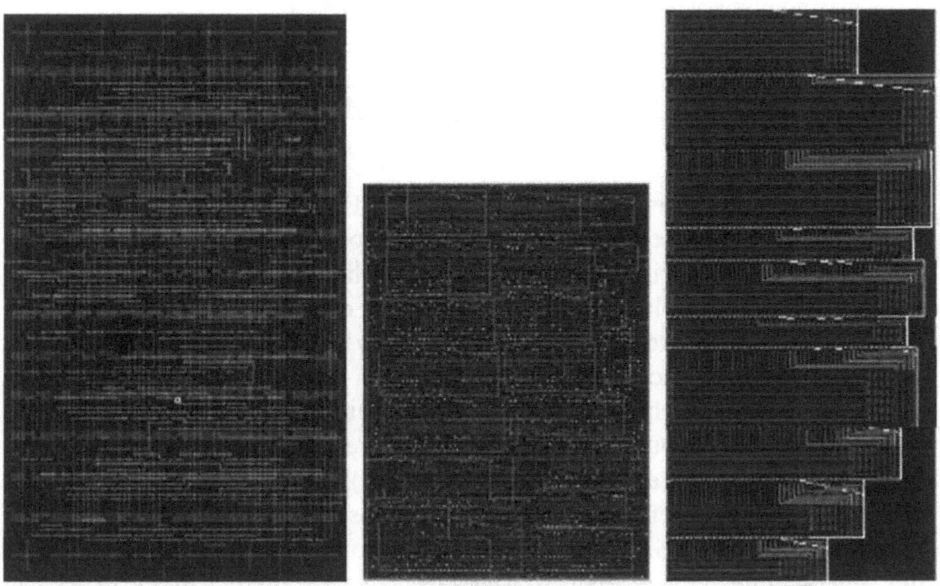

(a) SC-a (b) NPLA (c) RP-a

Figure 3.24: Layout of alu2 (n_L=10).

3.5.4.2. Glacier PLA

Three GPLA configurations are considered, named G1K, G5K and G10K respectively. The three FPGAs used for comparison are the *Xilinx* XC4000XL series, 4002, 4020, and 4062, which have similar numbers of programmable bits as G1K, G5K and G10K respectively. Some important parameters are listed in Table 3.8 [XILI99]. The same LGSynth 91 benchmark examples were tested. The GPLA synthesis algorithm is written in JAVA, except the multi-level logic synthesis uses *SIS* [SENT92] and the SOP minimization uses *ESPRESSO* [RUDE87]. The FPGAs are synthesized by *Synplicity Synplify Pro* 6.2.4. Due to the lack of a technology library for the programmable devices, we do not have area and delay information available in the experiment. Only a qualitative measurement of "whether the circuit can be implemented in the configuration" is given.

Table 3.8: Configuration parameters

	GPLA			*Xilinx* FPGA		
	G1K	G5K	G10K	4002	4020	4062
max.no.primary inputs	85	185	384	64	224	384
max.no.primary outputs	214	540	1132			
max. no. levels	12	18	24			
no. programmable bits	54k	456k	1.41M	61k	522k	1.43M

The results are shown in Table 3.9. For the FPGA implementations, some of the failures are caused by insufficient I/Os, not insufficient logic resources. If I/O limit is not an issue, the results show that GPLAs and FPGAs have similar powers for implementing the benchmark circuits.

We ware not able to compare area and delay of GPLA and FPGA, lacking of area and timing model of FPGA. The area and delay characteristics of GPLA are similar to those of RPLA, since their structures are basically the same except the layout patterns of GPLA are fixed (this is called a via programmable structure, detailed in Section 3.8); so the area/delay results for RPLA and SC can be used as a rough comparison for GPLA and SC. It has been reported that FPGA can be several times to an order of magnitude larger and several times slower than SC [SMIT97]. So we predict that GPLAs can be a viable alternative to a widely used FPGA structure.

Table 3.9: Comparison of GPLA and FPGA
(● means a successful mapping; and ○ means that the logic can be fully mapped, but the I/O ports are insufficient)

circuit	# PI	# PO	G1K	G5K	G10K	4002	4020	4062
apex7	49	37	●	●	●	○	●	●
alu2	10	6	●	●	●		●	●
C1355	41	32	●	●	●	○	●	●
apex6	135	99		●	●		○	●
alu4	14	8		●	●		●	●
x3	135	99		●	●		○	●
C2670	234	139			●		○	●
k2	45	43		●	●		●	●
C3540	50	22		●	●		●	●
dalu	75	16		●	●		●	●
i8	133	81		●	●		●	●
i10	257	224			●		○	○
C5315	178	123		●	●		○	●
C6288	32	32			●		●	●
C7552	206	107			●		○	●
des	256	245			●		○	○

3.5.5. Summary

RPLAs and their design methodology are described. The high regularity of the RPLAs is attractive in DSM integrated circuit design, since it greatly relieves the burden of analyzing huge amounts of layout patterns. More importantly, RPLAs have a simple design methodology; conventional design flow steps of technology mapping, placement and routing are eliminated. Thus there is no outer loop in the design for timing closure and hence the design cycle is shortened. Experimental results show that, compared to other structures, RPLAs provide similar delays. The RPLAs implemented with 2 metal layers consume on average 23% more area than standard-cells with 3 layers, however on average 22% of the area overhead is white space, which might be utilized by other circuits. RPLAs offer a good alternative to the standard-cell structure, providing higher circuit regularity and easier design methodology.

The GPLA is a re-configurable version of the RPLA. The contents of the PLAs are programmable, but all PLA sizes and interconnections are fixed. The design methodology is also simple; no technology mapping, placement and routing are needed.

Extension to sequential circuits is easy for RPLAs and GPLAs. The space between PLAs, under the river routing wires, can be used to build registers. One more metal layer (third layer) can be used to route feedback wires from higher level PLAs to lower level PLAs, and these can also be completed via river routing.

3.6. Whirlpool PLA

A Whirlpool PLA (WPLA) is a cyclic four-level programmable array. Since this cascaded NOR structure allows binary inputs to each plane, it extends the conventional SOP form. An algorithm called *Doppio-ESPRESSO* is developed to synthesize logic into WPLAs. Unlike *ESPRESSO* [RUDE87], which could be used to minimize two two-level circuits separately, *Doppio-ESPRESSO* uses the extra structural flexibility in WPLAs for further improvement. An important feature is that after logic synthesis, the layout is completely determined. No technology mapping, placement or routing is needed for the WPLA; neither is prediction necessary, because area and delay are solely determined by the logic embedded in the WPLA. Another interesting feature is that area and delay reductions rarely conflict in this structure (primarily because 4-level logic is required). The WPLA structure is suitable for circuits with up to several thousand gates.

The discussion of the WPLA is organized as follows. In 3.6.1, the WPLA structure is described, and area and delay computations are given. In Section 3.6.2, a synthesis algorithm for the WPLA structure is detailed. Section 3.6.3 gives experimental results, and Section 3.6.4 concludes.

3.6.1. The circuit structure of the WPLA

The WPLA structure is shown in Figure 3.25. The four programmable planes, labeled 0, 1, 2 and 3, are organized in a cycle. In each plane, input signals consist of external inputs as well as outputs from the preceding plane. Placing registers between planes 3 and 0 breaks the combinational loops. A plane is logically one level of NOR gates. Positive and/or negative (inverters) buffers are inserted between two neighboring planes; hence inputs to a NOR plane can have both polarities. WPLA circuits consume only 2 metal layers.

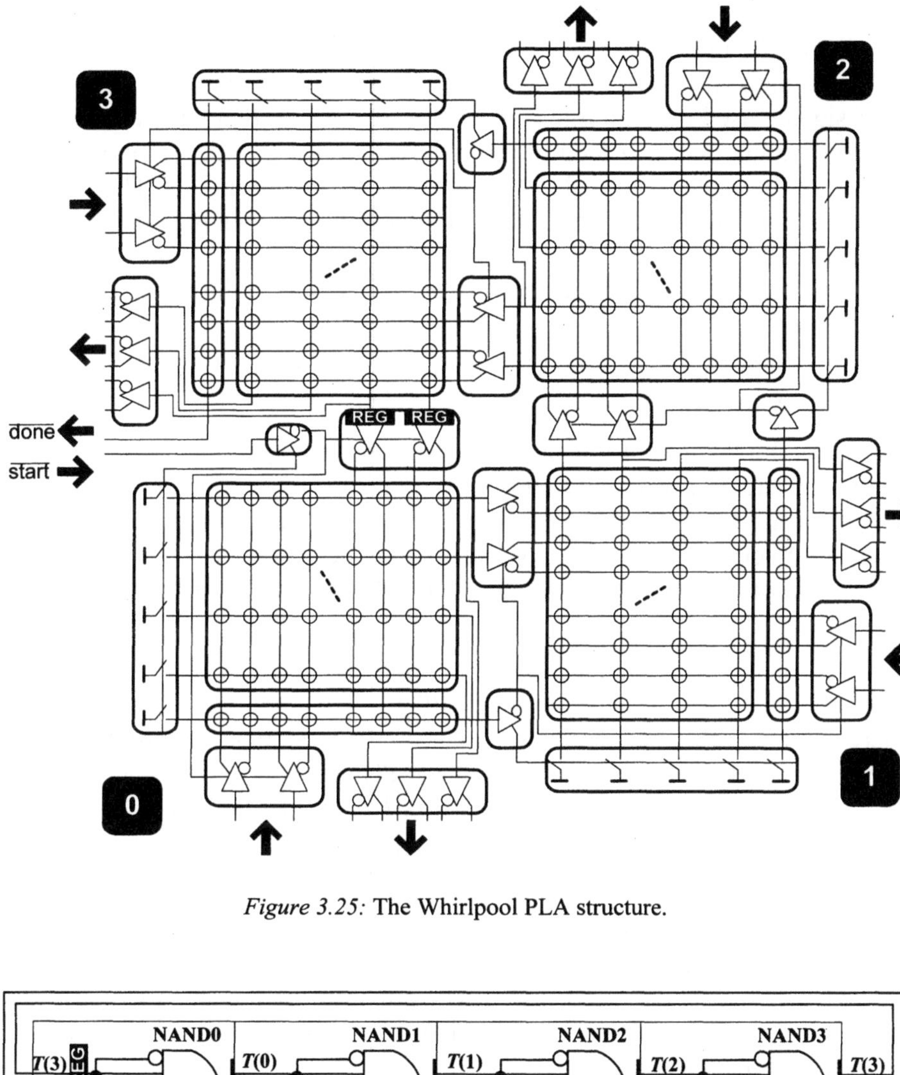

Figure 3.25: The Whirlpool PLA structure.

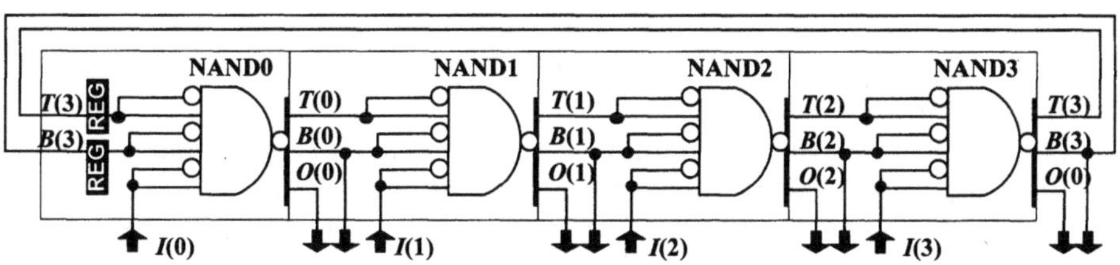

Figure 3.26: Abstract view of a WPLA.

Two cascaded NOR gates, together with the buffers, can be modeled as a NAND-NAND structure, shown in Figure 3.26, which is equivalent to a SOP. However, the inputs to the NAND gates can have both polarities available by choosing buffer polarities.

The signals of a WPLA are divided into four categories. *T*(.) denotes the set of signals used only internally by the next NAND. *B*(.) denotes the set of primary outputs that also feed the next NAND. *O*(.) are the primary outputs that do not fanout to the next NAND. *I*(.) are the primary inputs. The union of *T*(.) and *B*(.) is abbreviated by *TB*(.). Similar abbreviations include *BO*(.) and *TBO*(.).

Based on the area formulation of the NOR plane as described in Section 3.3, the width and height are

$$W(n) = |I(n)|u_I(n) + |B(n-1)|u_B(n-1) + |T(n-1)|u_T(n-1)$$
$$H(n) = |T(n)|u_T(n) + |B(n)|u_B(n) + |O(n)| + W_{BUFI},$$

where |.| is the number of the signals in the type, W_{BUFI} is the size of the input buffer, and *u*(.) is a variable denoting the ratio of unate signals of a type. If only one polarity of a signal is used in the plane, then this signal is classified as unate. Otherwise it is binate. A unate signal occupies one line in the plane, while a binate signal occupies two. So the binate coefficient of *B*(.) type is defined as

$$u_B(.) = \frac{|B_{una}(.)| + 2 \times \left[|B(.)| - |B_{una}(.)|\right]}{|B(.)|},$$

where $B_{una}(.)$ is the set of unate signals in *B*(.). Similar definitions can be derived for $u_T(.)$ and $u_I(.)$. The size of the WPLA is

$$W = \max[W(0), H(3)] + \max[H(1), W(2)] + W_{BUFM}$$
$$H = \max[W(3), H(0)] + \max[H(2), W(1)] + W_{REG},$$
$$Area = W \times H$$

where W_{BUFM} and W_{REG} are the size of the intermediate buffer and the register, respectively. Hence the area of the WPLA is completely determined by the embedded logic. The total delay is a summation of the delays of the four planes. The area and delay of a WPLA are fully determined by the logic functions it implements.

For the WPLA structure, reducing size can usually reduce delay. This can be explained by two factors: (1) the number of logic levels is fixed; and (2) uniform buffering is used in WPLAs. This characteristic makes the design flow straightforward; synthesis algorithms only need to focus on reducing the area since usually delay is reduced as well.

3.6.2. *Doppio-ESPRESSO*

The basic idea of WPLA synthesis is to reduce a pair of NANDs, and iterate for different pairs until no further improvement. The possible pairs in the WPLA are 0-1, 1-2 and 2-3. The synthesis of a pair of NANDs differs from conventional SOP synthesis since the WPLA structure allows negated products. To employ SOP synthesis, we transform form the NAND-NAND to SOP form, apply a SOP minimizer and transform the result back. The *Doppio-ESPRESSO* is summarized in the following pseudo code:

```
Doppio-ESPRESSO
do until no improvement {
    {nA,nB}=SOP2NN(ESPRESSO(NN2SOP(NAND0,NAND1)))
    if {nA,nB} better than {NAND0,NAND1}, {NAND0,NAND1}={nA,nB}
    {nA,nB}= SOP2NN(ESPRESSO(NN2SOP(NAND1,NAND2)))
    if {nA,nB} better than {NAND1,NAND2}, {NAND1,NAND2}={nA,nB}
    {nA,nB}= SOP2NN(ESPRESSO(NN2SOP(NAND2,NAND3)))
    if {nA,nB} better than {NAND2,NAND3}, {NAND2,NAND3}={nA,nB}
}
```

Here "improvement" and "better" mean smaller total area. Without loss of generality, we concentrate on NAND1-NAND2, and the labeling of the signals is shown in Figure 3.27.

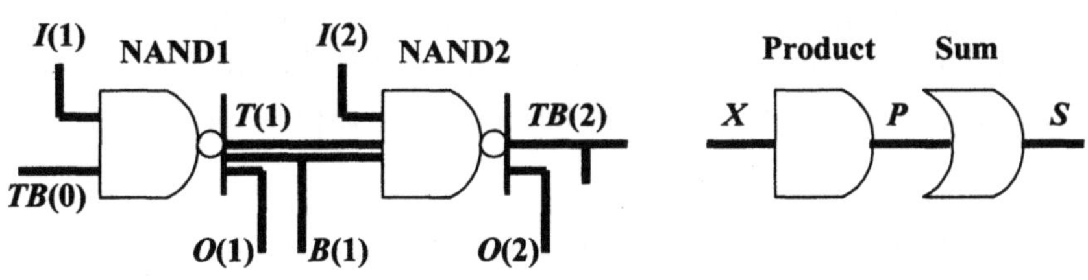

Figure 3.27: The signal labeling.

An example is used throughout the discussion of the algorithms. Following are the initial nand1-nand2 matrices,

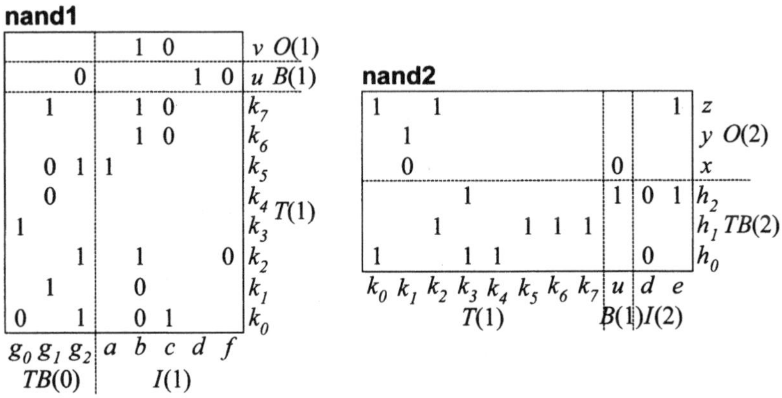

called nand1 and nand2. The inputs to nand1 include $TB(0)$ and $I(1)$, and the outputs of nand1 include $T(1)$, $B(1)$ and $O(1)$. $T(1)$, $B(1)$ and $I(2)$ form the inputs to nand2, and nand2 outputs $TB(2)$ and $O(2)$. For better visualization, only the care bits are shown. The SOP form has a product matrix and a sum matrix. The vacant space in the product matrix means '-', while the vacant space in the sum matrices means '0', or "don't output". The conventions of the SOP form include:

(1) The use of negative products is forbidden.

(2) The products output to the sums only.

(3) The sums only take the products as inputs.

The transformation algorithm should generate and accept the SOP form with these restrictions. In addition, the polarities of the primary inputs and outputs can be obtained by using the appropriate input and output buffers. So if the original function outputs signal Z, it is possible to end up with a function of the complement $\overline{Z}$ which has a simpler expression.

3.6.2.1. *NN2SOP*, the NAND-NAND to SOP transformation

We start with the simplest case, that is, $I(2)=\Phi$, $BO(1)=\Phi$ and the nand2 matrix is positive unate. Then the transformation is simply copying the nand1 matrix to the product matrix and copying the transpose (a 90 degree clockwise rotation followed by a mirroring of the X-axis) of the nand2 matrix to the sum matrix. Suppose $BO(1)\neq\Phi$. This is one of the major structural differences between WPLAs and conventional PLAs. Since the SOP form can only output from the sums, the *BO*(1) signals have to be raised to *O*(2). Suppose *Y* is a *BO*(1) signal. We can create a new *O*(2) signal, $\overline{Y}$, which is simply the complement of *Y*, but now corresponds to an output of the sum. Another case is *I*(2)≠Φ, which means some external signals enter nand2 directly. Since only product terms can enter the sums in SOP form, the *I*(2) signals need need to be pushed back to *I*(1):

$$Z = \overline{\prod_i Y_i \prod_j y_j} = \overline{\prod_i Y_i \prod_j \overline{T_j}}$$

where *Z* is an *O*(2) signal, y_j's belong to *I*(2), and $T_j = \overline{y}_j$, for all *j*, are the new *T*(1) signals. They replace *I*(2) and can be used in nand2. Finally, nand2 is not positively unate (it is always so in a SOP), which means nand2 can use both polarities of its input signals. Suppose an *O*(2) signal *Z* is the NAND of a set of positive *T*(1) signals Y_i and a set of negative *T*(1) signals Y_j. Each Y_j can be expressed by a NAND of a set of *TB*(0)∪*I*(1) signals x_{jk}. Then replace Y_j by x_{jk}'s:

$$Z = \overline{\prod_i Y_i \prod_j \overline{Y_j}} = \overline{\prod_i Y_i \prod_j \overline{\overline{\Pi x_{jk}}}} = \overline{\prod_i Y_i \prod_{j,k} \Pi x_{jk}} = \overline{\prod_i Y_i \prod_{j,k} x_{jk}} = \overline{\prod_i Y_i \prod_{j,k} \overline{\overline{x}_{jk}}}$$

However as mentioned above, x_{jk}'s have to be relayed to *T*(1); thus their polarities should be adjusted. Now the steps of the transformation to SOP form are described.

Step 1: The inputs of the product matrix include *TB*(0) and *I*(1)∪*I*(2). The outputs of the sum matrix include *TB*(2), *O*(2) and $\overline{BO(1)}$. Here $\overline{BO(1)}$ means the set of complemented *BO*(1) signals. We use similar abbreviations for other signals.

Step 2: Copy nand1 to the product matrix.

Step 3: For each input signal in the product matrix, build two rows, one with a '0' in the column of that signal, and one with a '1' in the column. By doing so, all the inputs of the product matrix, including *TB*(0) and *I*(1)∪*I*(2), are relayed. The relayed signals appear as if they are new *T*(1) signals that can be used by nand2. We call them pseudo-*T*(1) signals.

Step 4: Copy the transpose of nand2 to the sum matrix. Use the pseudo-*T*(1) signal when *I*(2) is required. For example, $h_0 = \overline{k_0 k_3 k_4 \overline{d}}$ where *d* is an *I*(2) signal. Then the pseudo *T*(1) signal *d* is used instead. Thus the 0's in the sum matrix come only from the *TB*(1) part of nand2, $x = \overline{\overline{k}_1 \overline{u}}$, for instance. The above equation is used to break the negative *TB*(1) literals into a set of *TB*(0)∪*I*(1) signals, which have their pseudo-*T*(1) versions available.

Step 5: Use a column singleton to relay a $BO(1)$ signal, such as $\bar{u}$ and $\bar{v}$. Notice that if the net is $\overline{O(1)}$, then it is also a row singleton in the sum matrix, because no $TBO(2)$ signal uses it. But $\overline{B(1)}$ must not be row singleton, because by definition some of the $TBO(2)$ use it.

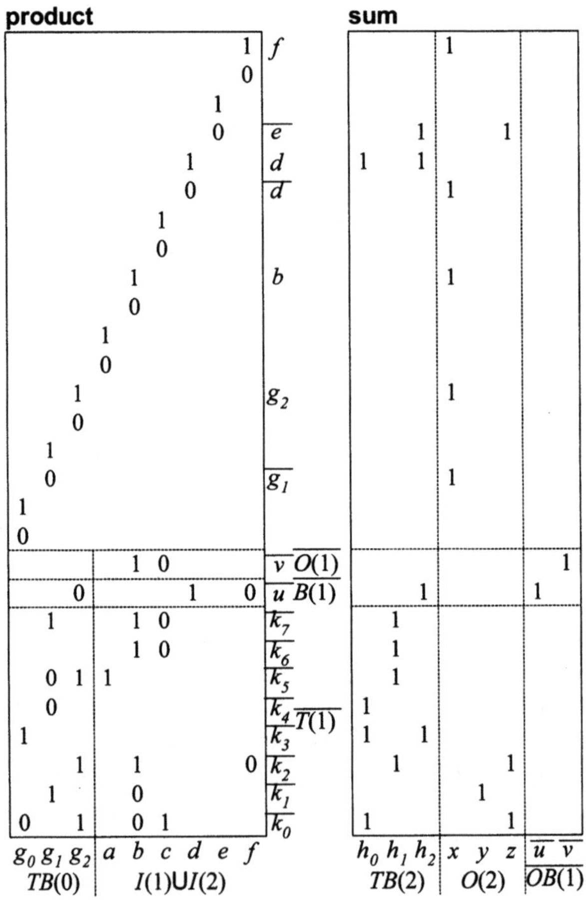

It is obvious that all the pseudo-$T(1)$ signals are not always utilized. But keeping them does not affect the SOP operation. A more succinct SOP representation, with all unused pseudo-$T(1)$ rows removed, looks like:

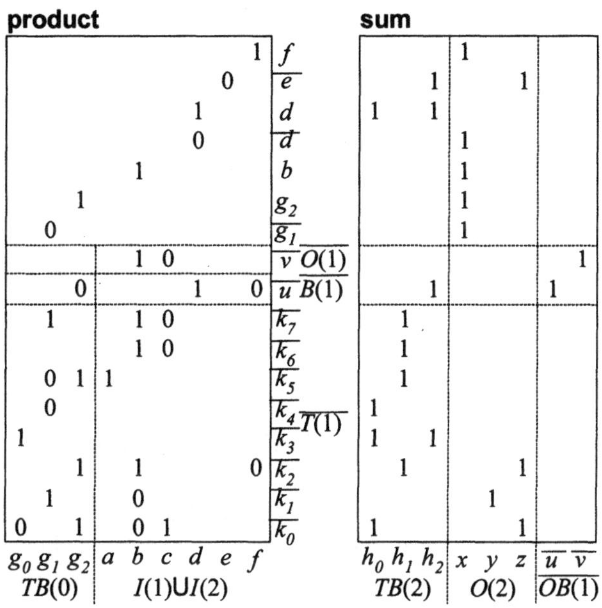

3.6.2.2. *ESPRESSO*, the SOP minimization algorithm

ESPRESSO is employed to perform the SOP minimization [RUDE88, BRAY90]. It makes no change to the input net list and the output net list. However, the content of the product and sum matrices may change. In the example, *ESPRESSO* gives the following SOP form.

product

g_0	g_1	g_2	a	b	c	d	e	f
								1
						0		
				1				
		1						
				1	0			
	0	1	1					
						1		
	0							
							0	
1								
		1		1				0
	1			0				
		0				1		0
0		1		0	1			

$TB(0)$ = $g_0\, g_1\, g_2$

sum

h_0	h_1	h_2	x	y	z	$\overline{u}$	$\overline{v}$
			1				
			1				
			1				
			1				
	1						1
	1						
1		1					
1			1				
		1			1		
1		1					
	1				1		
				1			
						1	
1					1		

$TB(2)$ = $h_0\, h_1\, h_2$

3.6.2.3. *SOP2NN*, the SOP to NAND-NAND transformation

During the transformation from the SOP back to a nand1-nand2 form, the original *BO*(1) and *O*(2) may have a new distribution between the two NANDs. The external inputs *I*(1) and *I*(2) may have new distributions as well. We will show that the re-distribution provides additional opportunities to reduce the logic functions. However, *TB*(0) and *TB*(2) are unchanged, because these signals are fixed, due to the structural restriction of the WPLA.

The algorithm consists of two parts. The first part, Steps 1 to 4, produces the nand1 and nand2 matrices from the SOP. These steps are mostly done by definition. Then the nand1-nand2 is further reduced using Steps 5 to 7.

Step 1: This step distinguishes $\overline{B(1)}$, $\overline{O(1)}$ and *O*(2). In the sum matrix, the *TB*(2) columns are left alone. In the remaining columns, the ones with a single 1 become $\overline{BO(1)}$, because these columns correspond to relays. The others are *O*(2). Shade the $\overline{BO(1)}$ columns. Then check the rows associated with the $\overline{BO(1)}$ column singletons. If a '1' is also a row singleton, then the associated $\overline{BO(1)}$ signal should be $\overline{O(1)}$. Otherwise it is $\overline{B(1)}$. Label the signal types of the recognized $\overline{B(1)}$ and $\overline{O(1)}$, and obtain the following SOP matrices.

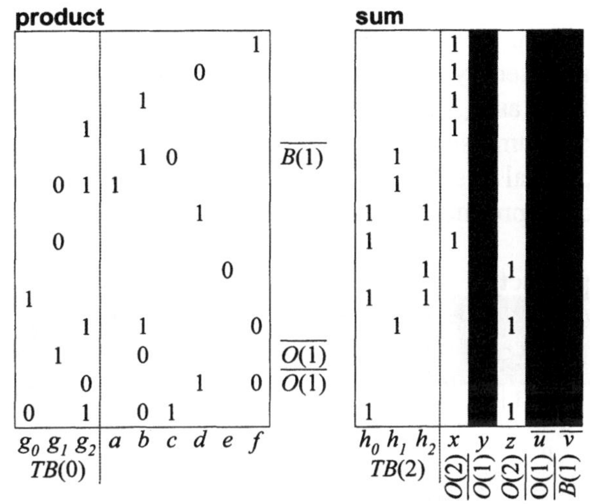

Step 2: This step recognizes $I(2)$ and $\overline{T(1)}$. In the product matrix, leave alone the $\overline{BO(1)}$ rows. Check the remaining rows. If a row has care bit(s) in *TB*(0) columns, or it has two or more care bits in the row, then the row is associated with a $\overline{T(1)}$ signal. The remaining rows, with single '0' or '1' in the non-*TB*(0) columns are *I*(2). Shade these rows, and label the corresponding columns with *I*(2).

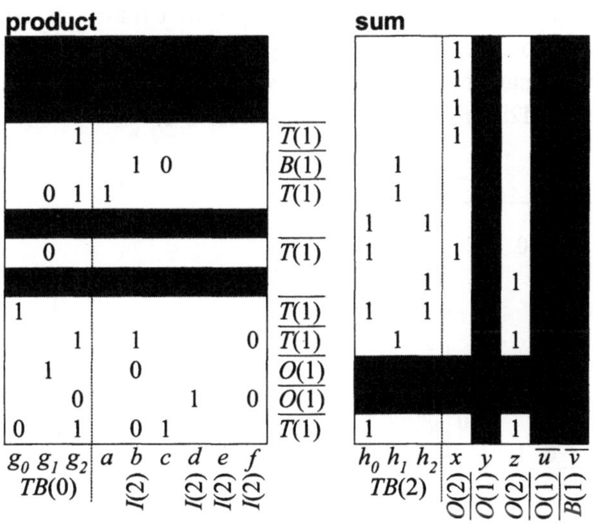

Step 3: This step identifies $I(2)$-only signals, because some $I(2)$ can also be $I(1)$, if they are used by both the nand1 and nand2. Check each column that has been identified as $I(2)$. An $I(2)$-only signal requires that each care bit appearing in the column should be the row singleton. Otherwise it is also $I(1)$. In the example, signal *e* is identified as an $I(2)$-only signal. Shade all the $I(2)$-only columns in the product matrix.

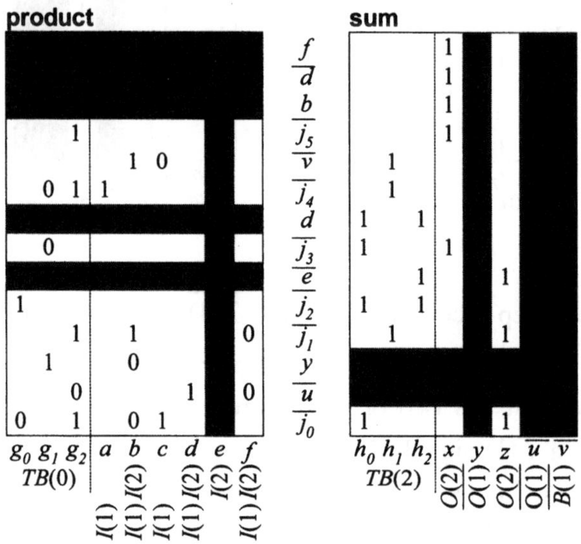

Step 4: The non-shaded region in the product matrix is copied to nand1, and the transpose of the non-shaded region in the sum matrix is copied to nand2, but the $I(2)$ care bits should be inverted. After the operation, the $TB(1)$ columns in the nand2 matrix should contain no 0's.

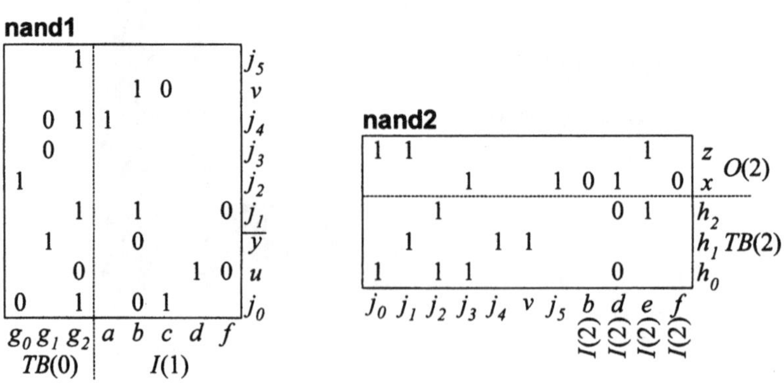

Re-arranging rows and columns, we get:

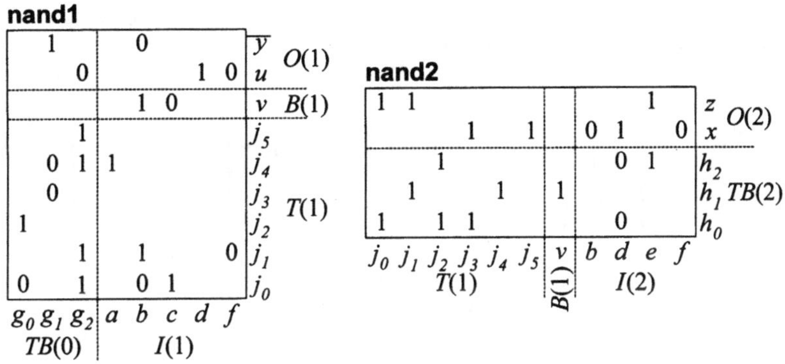

So far, the gain comes only from the SOP transformation. Further reduction is possible. Suppose:

$$Z = \overline{\prod_i Y_i \prod_j T_j},$$

where, $T_j = \overline{x}_j$, is a single-literal function in the nand1 and the literal x_j is a $TB(0)$. Then all T_j's can be combined in nand1 using a new $T(1)$ signal N:

$$N = \overline{\prod_j x_j},$$

which leads to:

$$Z = \overline{\prod_i Y_i \overline{N}}.$$

Now we have the option of removing some care bits in nand2 and replacing them with a new signal. If enough care bits are removed from the nand2 matrix by applying this transformation, then some columns might become empty and thus can be deleted. The cost is to introduce new columns for signals like N in the above formulas.

The transformation is illustrated by the simple example of Figure 3.28. Signal a and b are $TB(0)$ signals so should always stay at the input of nand1. Signal $u1$ and $u2$ are signals not touched. Signal i belongs to $I(1) \cap I(2)$, which means it is used by other functions in nand1 (although not shown in the figure). *ESPRESSO* can do nothing in this example; thus the first four steps in the *SOP-to-NN* will return the original structure. However, it is possible to organize a, b and i into one new nand1 term and make use of it in nand2. The transformation saves one *TB*(1) term. The following steps implement this idea.

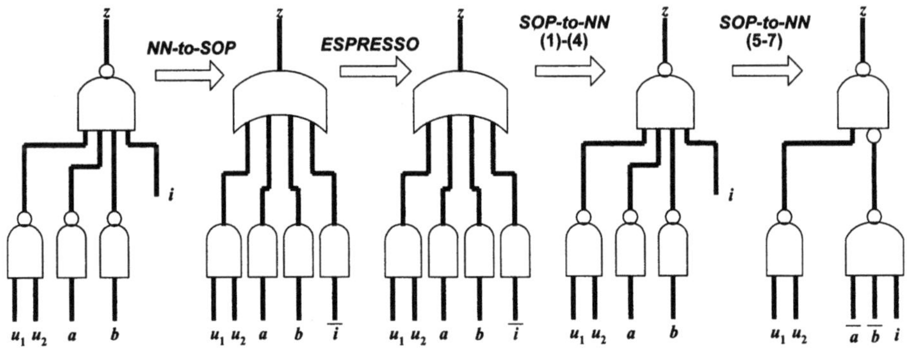

Figure 3.28: An example of the *SOP-to-NN* transformation.

Step 5: First, search for qualified *TB*(1) signals in the nand1 matrix, i.e., row singleton with the single care bit in a *TB*(0) columns. Shade the columns in the nand2 matrix associated with the selected *TB*(1) signals. Also in the nand2 matrix, shade the *I*(2) columns except the *I*(2)-only columns.

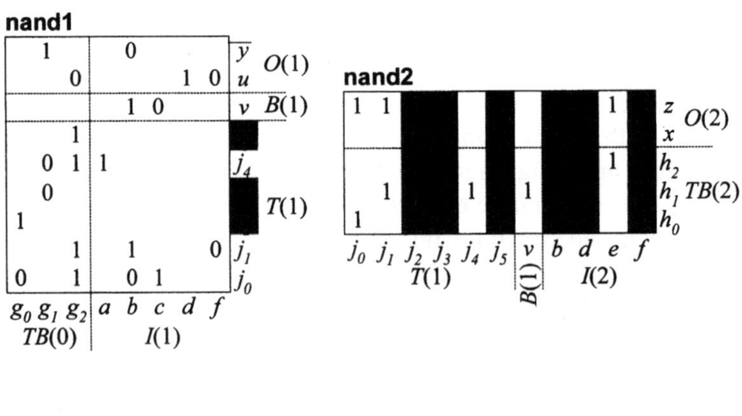

All the shaded columns in nand2 form a sub-matrix *S*.

S

$T(1)$: j_2	j_3	j_5	$I(2)$: b	d	f	
						z
	1	1	0	1	0	x
1				0		h_2
						h_1
1	1			0		h_0

Step 6: We identify common patterns in the S matrix. To eliminate a column, all the care bits in the column should be covered by some selected pattern(s). Denote the number of $T(1)$ columns eliminated by C_T, and the number of $I(2)$ columns eliminated by C_I. Eliminating these columns will save C_T+C_I columns in nand2 and C_T rows in nand1, at the expense of creating R_P new rows in nand1 and R_P new columns in nand2. Here R_P is the number of patterns used. Define *gain* as the total reduction in size of the two matrices:

$$gain = (C_T + C_I - R_P)H_2 + (C_T - R_P)W_1,$$

where H_2 is the height of the nand2 matrix, and W_1 is the width of the nand1 matrix. To simply the pattern recognition when different polarities may exist in the same signal, the S matrix is expressed in a pattern matrix S_P as shown below.

S_P

$T(1)$: $\bar{j}_2$	j_2	$\bar{j}_3$	j_3	$\bar{j}_5$	j_5	$I(2)$: $\bar{b}$	b	$\bar{d}$	d	$\bar{f}$	f	
												z
			*		*	*			*	*		x
	*							*				h_2
												h_1
	*		*					*				h_0

Each column in S is split into two, one for the positive literal, and one for the negative literal. Then the care bits are replaced by *'s. The algorithm has two parts. The first collects a set of candidate patterns for the covering, and the second selects a subset that maximizes the gain. The first part is summarized below.

```
CollectSetOfCandidatPatterns
all the *'s are labeled "uncovered"
PATTERN=Φ
for each column c {
      temp pattern p=*'s in c
      hasCommon=false
      for each column cc except c{
            if p⊆cc{
                  the *'s of p∩cc are labeled "covered"
                  hasCommon=true
            }
      }
      if hasCommon=true {
            the *'s in c are labeled "covered"
            PATTERN∪=p
      }
}
```

After the first step, the set *PATTERN* contains the candidate patterns. Define the size of a pattern as the product of the number of *'s in the column and the number of columns in matrix S_P that are covered by this pattern. Then a seed pattern, p_0, is chosen from the candidate set, which gives the highest gain. Notice that the highest gain provided by p_0 alone might not be positive, because $R_P = 1$, while C_T and C_I might both be 0 at this moment. If a tie occurs, choose the larger pattern. Further ties can be broken by choosing the one with the larger number of *'s in the column. The second part of the algorithm is as follows.

```
SelectPatterns
select the seed pattern p0
list[0]= p0
pattern-= p0
get gain[0]
i=1
while PATTERN≠Φ {
      choose p from PATTERN that increases gain the most,
            or, if none exists, choose the one decreases gain the least.
      list[i]=p, PATTERN−=p
      get gain[i]
      i++
}
find the maximum gain[n]. If tie, use the first one.
if gain[n]≤0, choose nothing; else choose the first n patterns in list.
```

In this example, two patterns are chosen as shown below.

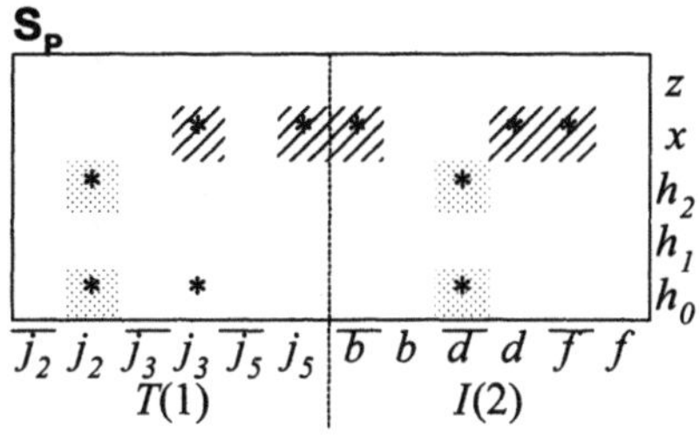

Remove the columns in nand2 covered by the chosen patterns, and replace their functions with new $T(1)$ signals. In the example, column j_2, j_5, d, b and f in nand2 are removed, and m_0 and m_1 are created.

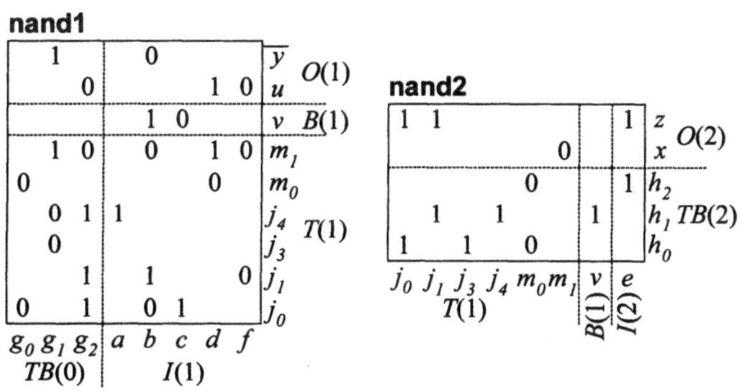

Step 7: Check if an $O(2)$ signal now becomes a row singleton in the nand2 matrix. For instance, $O(2)$ signal x now has only a single care bit in the row, which means that it can be pulled back to nand1 and become a $O(1)$. When the row is saved, it might generate empty columns in the nand2 matrix, m_1 in this case. Then the m_1 column can be saved. This operation concludes the *SOP2NN* algorithm; the final nand1-nand2 matrices are shown below.

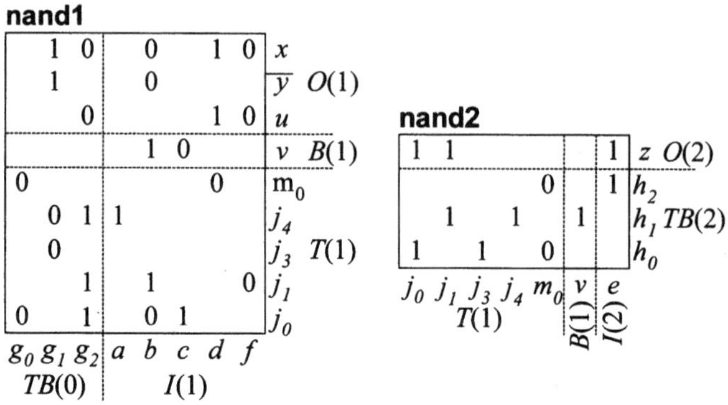

Finally we give the original logic functions:

$$k_0 = \overline{\bar{g}_0 g_2 \bar{b} c},\quad k_1 = \overline{g_1 \bar{b}},\quad k_2 = \overline{g_2 b f},\quad k_3 = \bar{g}_0,$$

$$k_4 = \overline{\bar{g}_1},\quad k_5 = \overline{\bar{g}_1 g_2 a},\quad k_6 = \overline{b\bar{c}},\quad k_7 = \overline{g_1 b \bar{c}},$$

$$u = \overline{\bar{g}_2 d \bar{f}},\qquad v = \overline{b\bar{c}},$$

$$h_0 = \overline{k_0 k_3 k_4 \bar{d}} = g_2 \bar{b} c + g_0 + \bar{g}_1 + d,\qquad h_1 = \overline{k_2 k_5 k_6 k_7} = g_2 b f + \bar{g}_1 g_2 a + b\bar{c},$$

$$h_2 = \overline{k_3 u \bar{d} e} = g_0 + d + \bar{e},\qquad x = \overline{k_1 \bar{u}} = \bar{g}_1 + b + g_2 + \bar{d} + f,$$

$$y = \overline{\bar{k}_1} = g_1 \bar{b},\qquad z = \overline{k_0 k_2 e} = \bar{g}_0 g_2 \bar{b} c + g_2 b \bar{f} + \bar{e}$$

and the final logic functions:

$$j_0 = \overline{\bar{g}_0 g_2 \bar{b} c},\quad j_1 = \overline{g_2 b f},\quad j_3 = \overline{\bar{g}_1},$$

$$j_4 = \overline{\bar{g}_1 g_2 a},\quad m_0 = \overline{\bar{g}_0 \bar{d}},$$

$$u = \overline{\bar{g}_2 d \bar{f}},\qquad v = \overline{b\bar{c}},$$

$$h_0 = \overline{j_0 j_3 \bar{m}_0} = g_2 \bar{b} c + \bar{g}_1 + g_0 + d,\qquad h_1 = \overline{j_1 j_4 v} = g_2 b f + \bar{g}_1 g_2 a + b\bar{c},$$

$$h_2 = \overline{\bar{m}_0 e} = g_0 + d + \bar{e},\qquad x = \overline{g_1 \bar{g}_2 \bar{b} d \bar{f}} = \bar{g}_1 + b + g_2 + \bar{d} + f,$$

$$\bar{y} = \overline{g_1 \bar{b}}_1,\qquad z = \overline{j_0 j_1 e} = \bar{g}_0 g_2 \bar{b} c + g_2 b \bar{f} + \bar{e}$$

Comparing u, v, h_0, h_1, h_2, x, y and z, the new functions are logically equivalent to the original ones, but the nand-nand size reduces from 8×10+11×6 to 8×9+7×4. *ESPRESSO* gives a nand2-nand2 with the size of 8×9+11×5. The additional improvement is due to the *SOP2NN* algorithm.

Following is a summary of the Whirlpool PLA structure. A WPLA is a four-level cyclic structure. The four programmable NOR arrays of a WPLA, labeled 0, 1, 2 and 3, are organized in a cycle. In each array, input signals consist of external inputs as well as outputs from the preceding array. Placing flip-flops between arrays 3 and 0 breaks combinational loops. All of the four arrays of the WPLA can output signals with both polarities. Only two metal layers are required to build a WPLA. The logical view of a WPLA is four cascaded NAND gates with binate inputs. The basic idea of WPLA synthesis is to reduce a pair of NANDs, and iterate for different pairs until no further improvement. In each iteration, a NAND-NAND is transformed to an SOP, an SOP minimizer is applied, and the result is transformed back.

3.6.3. Experimental results

We compare the following methods of implementation: standard-cells (SCs), network of PLAs (NPLAs) [KHAT00], River PLAs (RPLAs) [MO02b] and Whirlpool PLAs (WPLAs). An NPLA can be regarded as an intermediate representation between technology independent and technology dependent logic transformations [BRAY90]. The RPLA is a regular structure composed of a stack of PLAs; the adjacent PLAs are connected via river routing. Logically it represents a multi-level Boolean network. To eliminate confusion of the network level and logic level, we use the term "depth" to describe the network level. Since an SOP node in a network contains two logic levels (trivial nodes like AND, OR or buffer functions may contain less than 2 levels), the depth of a network is approximately half of the number of logic levels. In fact, a *depth*=2 NPLA or RPLA is logically similar to the WPLA, except that (1) WPLAs can have primary outputs directly from the product terms and (2) the product terms can appear in both polarities. The *depth*=1 NPLA and RPLA degrade to a single PLA.

A 0.35-micron technology was used for the comparisons. A standard-cell library with over 100 gates was available, and each logic gate has at least two choices of drive strength. Typical parameters of the gate library are given in Table 3.10.

Table 3.10: Typical parameters of the gate library

Parameter	ND2	ND2X4
logic function	2-input NAND	2-input NAND
area (um^2)	54	126
load limit	12	22
input pin load	1.0	1.2
intrinsic delay (ps)	170	540
load dependent delay (ps/load)	60	30

Standard-cell implementations use over-the-cell-routing. Since the gates use metal-1 for internal connections, metal-2 and -3 are needed for inter-gate connections. NPLAs use metal-1 and -2 for internal connections, so the NPLA needs metal-3 and -4 for inter-PLA connections. The RPLAs only need metal-1 and -2 for routing. A WPLA also uses only metal-1 and -2. Some typical parameters in the PLA designs are given in Table 3.11.

Table 3.11: Typical parameters in the PLA designs

Parameter	value
size of a programmable bit (um)	0.8x1.0
width of input/output buffer (um)	4.0
width of intermediate buffer (um)	10.0
intrinsic delay of a transistor (ps)	30
load dependent delay of a transistor (ps/loading transistor)	40
intrinsic delay of in/out buffer (ps)	250
load dependent delay of in/out buffer (ps/loading transistor)	8
intrinsic delay of intermediate buffer (ps)	150
load dependent delay of intermediate buffer (ps/loading transistor)	5

Fifteen FSM examples from the LGSynth 91 benchmark set [LGSY91] were tested. After the latches are removed from each example, (we do not deal with state minimization and encoding), the combinational part is processed with *SIS* [SENT92] (using *script.rugged*) to achieve an initial Boolean network with depth d_0. Then for SC, NPLA and RPLA synthesis, we generate area/delay trade-off curves, by decreasing the depth gradually from d_0 using the *SIS* command "reduce_depth –d *d*". Note that the command may not always reduce the depth to the designated value *d*, but to some value no greater than *d*. At each depth *d*, for SC we use *SIS* "map –n1 –AFG" command (delay-prior circuit that respects load limit) for technology mapping; for NPLAs, we cluster all single-output nodes at the same level, and call *ESPRESSO* with its default settings to minimize the clustered multiple-output PLAs. The RPLAs are synthesized with its own algorithm [MO02b]. WPLAs have a fixed depth of 2, so it only has one solution per example (no area-delay trand-off).

In Table 3.12, the number of programmable bits of NPLAs, RPLAs (both *depth*=2) and WPLAs are compared. In this case, both the NPLA and RPLA are four-level structures. However due to the different algorithms used to synthesize them, their results are slightly different. The differences in the bit numbers show the additional improvement achieved by the *Doppio-ESPRESSO* algorithm; *Doppio-ESPRESSO* achieves on average 20% more reduction than *ESPRESSO*. However, fewer programmable bits do not necessarily imply smaller areas, because PLA structures also contain components such as buffers etc. For WPLAs, there can be "white space" along the boundary and in the center, as illustrated in Figure 3.25.

Table 3.12: Bit counts of NPLA, RPLA (*depth*=2) and WPLA

circuit	NPLA	RPLA	WPLA
s208.1	1623	1609	1038
s298	4053	4133	3800
s344	7559	7559	6234
s349	7902	7819	6582
s382	7428	7508	7198
s386	10200	10200	8228
s400	6575	6616	5714
s420.1	4691	4728	4439
s444	7152	7288	7046
s526	8208	8208	7490
s641	21175	21175	16583
s820	17862	18840	13675
s838.1	37353	35759	28032
s1488	53958	54884	44211
s1494	56481	56070	47533
average	120%	120%	100%

Area and delay results are given in Table 3.13. The synthesis times are given in Table 3.14. The "depth" refers to the depth during the technology independent logic transformation. The values in the column are exact for the NPLA, RPLA and WPLA, since their logic levels will not change. However, for the SC, there is a technology-mapping step after the technology independent step, and the gate levels (including buffers when calculating levels) are shown in the "gate level" column. No placement or routing has been done for SCs and NPLAs, so these areas are just the raw areas of the logic components (100% area utilization). Although we can assume that routing is done on higher metal layers, in reality, SCs may need cap cells on both sides of the rows and feed-thru cells. NPLAs require block-level placement, which may generate white space. The RPLAs have their finalized layouts, which contain white space, so they give fair comparisons. In addition, the delays of the SCs and NPLAs may change after routing, due to parasitics on wires. In contrast, WPLAs consume no additional area nor have additional delay uncertainties.

Table 3.13: The comparison of area and delay

(part 1 of 2)

circuit	depth	gate level	area ($1000\mu m^2$) SC	NPLA	RPLA	WPLA	delay (ns) SC	NPLA	RPLA	WPLA
s208.1	8	9	4.91	1.72	8.82		3.12	3.91	3.92	
	4	9	5.26	1.70	8.10		3.02	2.33	2.49	
	3	9	5.94	1.82	7.86		2.47	1.92	2.02	
	2	9	5.98	2.22	8.36	2.46	2.47	1.70	1.74	1.89
	1	9	6.71	2.90	—		2.47	1.41	—	
s298	5	9	9.05	3.39	9.94		3.03	3.55	3.51	
	3	9	11.0	4.72	10.6		2.84	3.04	3.14	
	2	6	10.1	4.92	12.1	4.24	2.06	2.42	2.28	2.60
	1	6	11.3	3.73	—		2.06	2.48	—	
s344	9	13	9.83	5.08	8.81		3.94	5.92	5.76	
	5	11	14.8	7.06	8.46		3.56	4.41	4.33	
	4	11	15.5	5.84	10.1		3.21	3.69	3.65	
	3	10	14.1	7.62	10.6		2.78	3.56	3.50	
	2	11	18.4	8.57	10.1	8.44	3.12	3.10	3.09	3.32
	1	8	19.6	19.1	—		2.96	4.59	—	
s349	9	13	10.3	5.26	9.25		3.94	6.01	6.21	
	5	11	14.2	7.28	8.01		3.56	4.41	4.92	
	4	11	16.8	6.31	10.6		3.21	3.84	4.57	
	3	10	14.5	8.04	10.6		2.78	3.62	4.00	
	2	11	18.5	8.88	12.6	8.8	3.12	3.16	3.77	3.30
	1	8	19.7	19.1	—		2.96	4.59	—	
s382	8	13	12.6	4.14	12.6		3.56	5.23	5.52	
	4	11	15.5	5.60	11.6		3.12	3.81	4.02	
	3	10	15.4	6.61	10.6		2.82	3.47	3.76	
	2	9	17.1	8.43	13.6	10.1	2.68	3.10	3.50	3.38
s386	7	11	10.6	9.11	14.2		4.32	4.34	4.43	
	5	10	11.2	13.2	14.4		3.94	3.58	4.20	
	3	11	13.2	17.6	16.3		3.82	3.44	3.50	
	2	11	19.1	17.2	16.0	15.8	3.79	4.02	4.10	3.97
s400	8	12	12.7	4.64	10.5		3.88	5.38	5.38	
	4	10	12.8	5.42	10.1		3.17	3.66	3.70	
	3	8	12.7	6.34	9.96		2.93	3.28	3.31	
	2	10	14.9	7.44	9.20	8.0	2.46	2.88	2.99	3.11
	1	6	17.0	12.2	—		2.16	3.09	—	
s382	8	13	12.6	4.14	12.6		3.56	5.23	5.52	
	4	11	15.5	5.60	11.6		3.12	3.81	4.02	
	3	10	15.4	6.61	10.6		2.82	3.47	3.76	
	2	9	17.1	8.43	13.6	10.1	2.68	3.10	3.50	3.38
s386	7	11	10.6	9.11	14.2		4.32	4.34	4.43	
	5	10	11.2	13.2	14.4		3.94	3.58	4.20	
	3	11	13.2	17.6	16.3		3.82	3.44	3.50	
	2	11	19.1	17.2	16.0	15.8	3.79	4.02	4.10	3.97

(part 2 of 2)

circuit	depth	gate level	area (1000μm²) SC	NPLA	RPLA	WPLA	delay (ns) SC	NPLA	RPLA	WPLA
s400	8	12	12.7	4.64	10.5		3.88	5.38	5.38	
	4	10	12.8	5.42	10.1		3.17	3.66	3.70	
	3	8	12.7	6.34	9.96		2.93	3.28	3.31	
	2	10	14.9	7.44	9.20	8.0	2.46	2.88	2.99	3.11
	1	6	17.0	12.2	—		2.16	3.09	—	
s420.1	12	17	10.1	3.47	11.5		4.55	6.39	5.95	
	6	17	10.9	3.58	11.0		4.55	3.93	4.06	
	4	17	12.0	4.74	11.2		4.55	3.39	4.22	
	3	17	12.9	5.83	12.6		4.55	3.19	3.90	
	2	17	14.5	7.45	15.0	9.0	4.51	2.86	3.74	3.15
	1	17	15.7	14.6	—		4.51	3.50	—	
s444	8	13	12.1	4.41	11.5		3.98	5.38	5.53	
	4	9	13.5	5.23	10.1		3.22	3.66	4.61	
	3	8	13.7	6.13	10.0		2.82	3.22	4.20	
	2	9	16.4	8.10	12.1	9.0	2.54	3.04	3.09	3.23
	1	7	17.1	12.2	—		2.37	3.09	—	
s526	7	10	13.8	6.24	10.5		3.32	5.37	5.47	
	4	9	16.8	8.62	10.0		3.16	4.30	4.46	
	3	9	15.9	7.84	13.5		2.76	3.47	3.75	
	2	8	16.3	9.08	18.8	9.7	2.46	3.16	3.31	3.33
	1	9	19.5	10.5	—		2.29	2.73	—	
s641	14	18	13.1	5.94	12.7		5.29	8.33	8.30	
	7	16	19.5	8.19	13.0		4.82	6.22	7.75	
	5	15	22.2	10.5	13.4		4.39	5.62	7.86	
	3	13	29.5	13.9	18.6		4.47	5.70	6.98	
	2	12	35.8	14.0	25.4	14.9	4.04	4.81	7.06	4.41
s820	7	11	23.3	11.9	20.3		4.96	7.18	7.21	
	4	10	25.6	13.9	22.1		4.89	5.69	5.93	
	3	10	24.9	14.8	25.7		4.33	4.89	5.03	
	2	11	28.7	18.1	29.9	15.3	3.95	4.30	4.42	4.28
s838.1	20	21	23.0	18.6	33.5		8.67	11.5	10.0	
	10	21	24.1	19.1	39.9		8.67	7.32	9.91	
	7	21	28.1	21.8	44.1		8.67	6.70	10.4	
	5	21	27.2	34.0	49.8		7.41	6.13	8.02	
	4	21	30.7	37.1	55.2		6.92	5.98	6.79	
	3	33	33.4	43.3	55.8		7.02	6.10	5.81	
	2	33	36.1	45.8	54.1	33.2	6.51	6.77	5.92	7.16
	1	33	47.1	90.5	—		6.30	9.70	—	
s1488	5	11	46.5	38.2	59.2		7.04	9.62	9.21	
	3	11	49.9	44.5	60.3		6.96	7.68	7.70	
	2	11	54.2	50.5	65.3	54.2	6.87	7.15	6.96	6.94
s1494	5	11	47.0	38.4	50.2		7.11	9.67	9.58	
	4	11	50.5	43.1	55.1		7.02	8.66	8.47	
	3	11	48.2	38.8	67.2		6.92	7.51	77.9	
	2	10	54.6	52.6	70.1	54.5	7.05	7.21	7.31	7.03

Table 3.14: The comparison of synthesis time (minutes)

(part 1 of 2)

circuit	depth	level	SC	NPLA	RPLA	WPLA
s208.1	8	9	0.1	0.1	1.2	
	4	9	0.1	0.1	1.3	
	3	9	0.2	0.1	1.2	
	2	9	0.2	0.1	1.2	0.1
	1	9	0.3	0.1		
s298	5	9	0.1	0.1	1.1	
	3	9	0.1	0.1	1.0	
	2	6	0.2	0.1	1.4	0.1
	1	6	0.2	0.1		
s344	9	13	0.1	0.1	1.2	
	5	11	0.1	0.1	1.2	
	4	11	0.1	0.1	1.3	
	3	10	0.2	0.1	1.2	
	2	11	0.2	0.1	1.2	0.1
	1	8	0.3	0.2	1.2	
s349	9	13	0.1	0.1	1.5	
	5	11	0.1	0.1	1.5	
	4	11	0.1	0.1	1.5	
	3	10	0.2	0.1	1.4	
	2	11	0.3	0.1	1.5	0.1
	1	8	0.4	0.2		
s382	8	13	0.2	0.1	1.5	
	4	11	0.2	0.1	1.5	
	3	10	0.3	0.1	1.5	
	2	9	0.4	0.1	1.5	0.1
s386	7	11	0.1	0.1	1.4	
	5	10	0.2	0.1	1.4	
	3	11	0.3	0.1	1.4	
	2	11	0.4	0.1	1.5	0.1
s400	8	12	0.1	0.1	1.4	
	4	10	0.2	0.1	1.4	
	3	8	0.3	0.1	1.4	
	2	10	0.3	0.1	1.4	0.1
	1	6	0.5	0.3		
s420.1	12	17	0.1	0.1	1.6	
	6	17	0.2	0.1	1.6	
	4	17	0.2	0.1	1.7	
	3	17	0.3	0.1	1.7	
	2	17	0.4	0.1	1.7	0.1
	1	17	0.5	0.2		
s444	8	13	0.1	0.1	1.6	
	4	9	0.2	0.1	1.7	
	3	8	0.3	0.1	1.6	
	2	9	0.4	0.2	1.7	0.2
	1	7	0.4	0.2		

(part 2 of 2)

circuit	depth	level	SC	NPLA	RPLA	WPLA
s526	7	10	0.1	0.1	1.4	
	4	9	0.2	0.1	1.5	
	3	9	0.3	0.1	1.4	
	2	8	0.3	0.2	1.5	0.2
	1	9	0.5	0.2		
s641	14	18	0.1	0.1	1.3	
	7	16	0.2	0.2	1.3	
	5	15	0.3	0.2	1.4	
	3	13	0.5	0.3	1.3	
	2	12	0.6	0.3	1.3	0.4
s820	7	11	0.3	0.2	1.4	
	4	10	0.3	0.3	1.5	
	3	10	0.5	0.4	1.5	
	2	11	0.5	0.5	1.5	0.4
s838.1	20	21	0.3	0.2	1.5	
	10	21	0.3	0.2	1.5	
	7	21	0.3	0.2	1.5	
	5	21	0.3	0.2	1.5	
	4	21	0.3	0.3	1.6	
	3	33	0.5	0.4	1.5	
	2	33	0.7	0.6	1.5	0.6
	1	33	0.9	0.6		
s1488	5	11	0.5	0.5	2.7	
	3	11	1.6	0.7	2.8	
	2	11	2.0	0.8	2.8	0.9
s1494	5	11	0.8	0.5	2.9	
	4	11	1.0	0.6	3.2	
	3	11	2.4	0.8	3.3	
	2	10	3.2	0.9	3.3	0.9

For a better view of the experimental results, the area/delay data of the SCs, NPLAs and RPLAs are normalized with respect to the WPLA results and plotted in Figure 3.29. The (1,1) point represents the WPLA single point for all examples. Connected points for SC, NPLA or RPLA represent area/delay trade-off curves for a single example. Figure 3.29 shows that SCs generally have larger areas than WPLAs, but can provide smaller delays if more area is allowed. NPLAs are just the opposite; they can provide smaller (raw) areas, but usually are slower. Comparing the *depth*=2 cases, on average, WPLAs are 37% and 0% smaller than the raw areas of SCs and NPLAs respectively, but only 5% and 3% slower than SCs and NPLAs. However, recall that the areas of SCs and NPLAs only account for raw logic components and use more metal layers. After placement, the areas of both are expected to grow, especially NPLAs. So in reality, these area/delay curves would shift to the right relative to the WPLA point. The WPLA is on average 56% smaller than the *depth*=2 RPLA and 13% faster than it. The RPLAs are not expected to be as useful in implementing small circuits such as those in this experiment [MO02b], because when the circuit is small or the depth is small, the river routing region may occupy a large portion of the entire RPLA area. A rough estimate of the river routing area can be obtained by the difference between the area of the RPLA and the NPLA. Comparing the area of SC, NPLA, RPLA and WPLA with similar delays (may have different depths), we find that WPLA is on average 19% larger than NPLA (raw area), 26% smaller than SC and 32% smaller than RPLA.

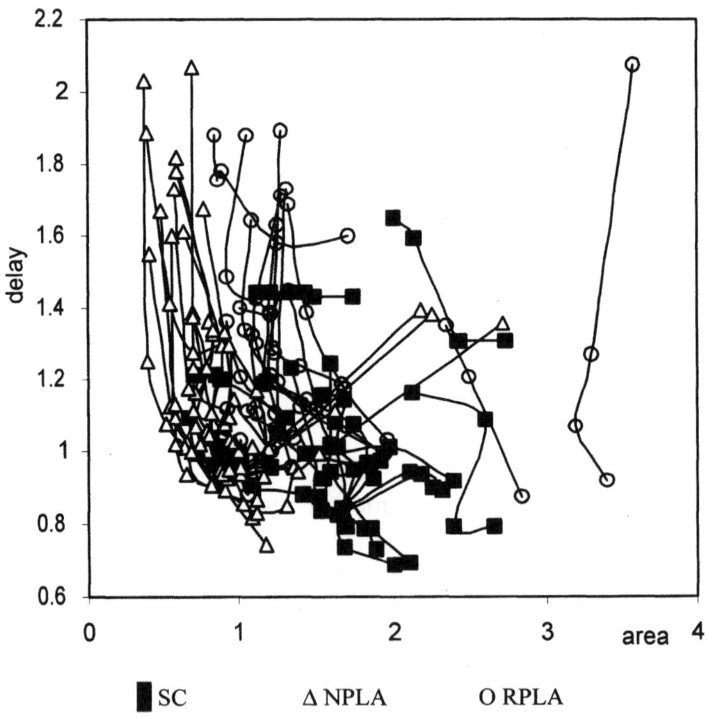

Figure 3.29: Normalized area/delay curves.

Note that some SC, NPLA and RPLA curves are not monotone decreasing with area; thus reducing the depth may not necessarily lead to faster circuits. Other curves are unpredictable in shape, so timing closure becomes even more difficult. Thus, in addition to uncertainty caused by physical design, area/delay relations of SCs, NPLAs and RPLAs are also unpredictable, while WPLAs do not suffer from such problems.

We also found that the number of gate levels after technology mapping is non-linear in the depth of the technology-independent result, and the relationship is not even monotonic. An interesting phenomenon is that in some circuits like "s838.1", when the depth is reduced, the actual number of gate levels increases. This can be explained by two factors. One is from the covering done in the technology mapping. Suppose the classical tree covering is used, where the technology-independent synthesis result is first transformed into a generic netlist with only NAND2's and inverters. If the depth is not small, the level of the generic netlist follows the depth quite well. But when the depth is very small, the nodes in the netlist are large, and many levels of NAND2's and inverters have to be used to represent them. This makes the levels of the generic netlist and thus the mapped gate netlist almost unpredictable. The other factor is the loading problem. As the depth goes down, it is conceivable that the loads (on nets between nodes, and the SOP connections within nodes) tend to increase. To obey the load limit and improve speed, appropriate buffering should be done during technology mapping, which also increases the levels of gates. This shows that even within logic synthesis, the technology-independent step has difficulty predicting the behavior of the technology-dependent step. The relationship between depth, gate levels, area and delay is complicated.

Logic synthesis times for NPLAs and WPLAs are usually smaller than SCs, because SCs need a technology mapping stage, which becomes notably slower as the circuit size increases. The RPLA synthesis times are the slowest, due to its iterative node-placement algorithm [MO02b].

3.6.4. Summary

Whirlpool PLAs (WPLAs) are logically four-level NOR networks. Their cyclic structure makes them compact. The design methodology for WPLAs involves only logic synthesis; no prediction is needed because area and delay are totally determined by the logic embedded in the WPLA. *Doppio-ESPRESSO*, a new four-level logic synthesis algorithm for WPLA synthesis exploits additional structural flexibility. Experimental results show that

WPLAs are quite competitive, in terms of area and delay, with standard cell implementations and network of PLA implementations, but are much more regular and predictable. It also is superior to another regular structure, the River PLA, in both area and delay, for the examples tested. A comparison between WPLAs and depth=2 NPLAs and RPLAs also shows the advantage of the *Doppio-ESPRESSO* algorithm in terms of the total number of programmable bits needed to build a circuit. However, some remaining problems require more discussion:

(1) The regularity of a chip involves both local and global regularity. The WPLA provides a structure with local regularity. However to integrate multiple WPLAs on a chip and achieve global regularity is not easy. The problem includes, partitioning of the circuit into many WPLAs, placing and routing them in a regular way.

(2) The pin positions of the WPLA are fixed after the synthesis. This seems worse than for SC implementations. However consider implementing the same logic functions (a part of a large circuit) with SC. The gates are usually placed closer, although not necessarily in a rectangular region. The pins connecting to the external circuit are actually on some of the gates. It is unlikely that these pins can be moved arbitrarily, because the gates need to maintain some spatial relations indicated by the gate-level placement. The pins can move within a small range by moving the gates carrying them. To move them farther, the only way is to flip the entire "SC block". However changing orientation of a "SC block" is not as flexible as a WPLA, because the "SC block" cannot do things like "rotate 90°". Therefore, the fixed pin position is not a serious drawback of the WPLA compared to the SC, because the WPLA can be thought of as "placed and routed".

(3) PLAs can also be re-sized to get different area/performance characteristics. Also the characterization of a set of PLA parameters is much faster than that of a library of hundreds of gates.

(4) Engineer Change Orders (ECOs) for SC implementations involve both synthesis and physical design modifications. But for the WPLA, it is mainly a synthesis problem.

(5) A programmable version of the WPLA is anticipated, and experiments need to be done to show if it is a good alternative to the LUT-based structures.

3.7. Checkerboard

The Checkerboard (CB) is an array of the so called Generalized OR (GOR) blocks. A GOR block is basically a variation of the dynamic NOR array, as described in Section 3.3. Within a GOR block, the lower layers, poly-silicon and metal 1, form a multiple output OR gate with selective input polarities. The higher layers, metal 2 and 3, implement the global interconnections of the GOR blocks. The GOR block works in a dynamic fashion, involving a pre-charging phase and an evaluation phase in one clock cycle, controlled by the "start" signal of the block. A CB is an array of the GOR blocks and register blocks. The blocks in the same column have the same width, and the blocks in the same row have the same height. The circuit is simplified and decomposed into a netlist of OR gates with inputs of both polarities. The netlist is levelized, and the gates on the same level are clustered into a set of GOR blocks. Hence blocks are labeled with levels. In one clock cycle, the evaluations of the blocks are done level by level.

The synthesis algorithm for CB does not require technology mapping, which saves significant time and manpower in migrating the methodology to a new process. The algorithm starts with a technology independent synthesis of a Boolean netlist. The result is decomposed into a network of OR gates with binate inputs and registers. Then the problem is to map this to a CB with $n_X \times n_Y$ blocks and design the global wires. There are two unique structural features. One is that gates in the same GOR-block with common inputs can share input pins, thereby reducing the number of connections. The other is that in each block, the input and output lines are orthogonal and these can be permuted arbitrarily and independently. The situation with the global wiring layers (metal2 and 3) is similar. Using these features, an integrated placement and routing algorithm is developed, adopting a simple spine net topology [MO03a]. This makes routing and timing highly predictable.

This section is organized as follows. In 3.7.1, the structure of the Checkerboard is described. The area and delay of the Checkerboard is formulated. The algorithm for Checkerboard is discussed in 3.7.2. Experimental results are given in 3.7.3, and 3.7.4 summarizes.

3.7.1. The structure of Checkerboard and its dynamic operation

The building block of the CB is the GOR-block, illustrated in Figure 3.30. The input signals enter from the left side, where their polarities are selected. They pass a switch before controlling the NMOS transistors that implement the logic functions. A switch is composed of a PMOS pass transistor and an

NMOS pull down transistor. When the “start” signal is low, the NMOS is turned on, so that it always outputs “low”; when “start” is high, the NMOS is off and the PMOS is transparent, so that the input signal (after polarity selection) can control the transistors that form the logic functions. A NOR function is one vertical line in the logic function block, controlled by the horizontal input signals if NMOS transistors are built at the cross points. The vertical line can be pre-charged to “high” by the PMOS transistors that are controlled by “start” as well. The pre-charging and evaluation are mutually exclusive, controlled by the “start” signal. The “start” signal enters at the left/top corner of the block. It first passes a delay adjusting buffer, which is used to compensate for skew due to the propagation delay. The vertical NOR lines are finally buffered by inverters and the output pins are at the top side.

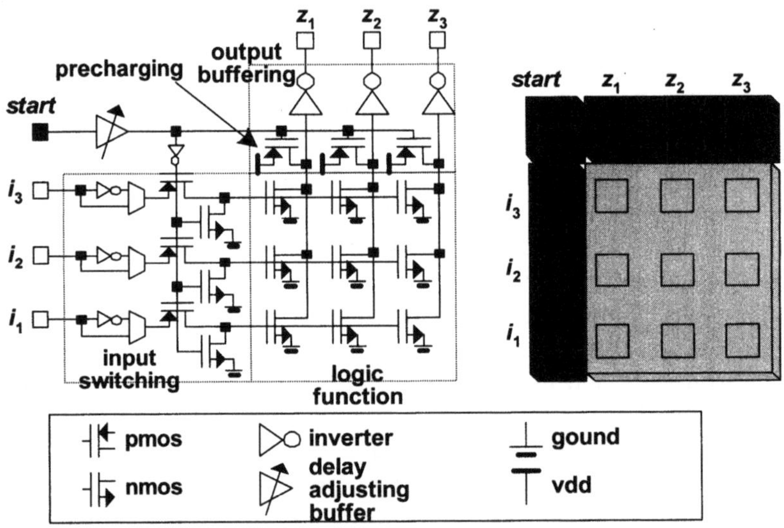

Figure 3.30: The GOR block of the Checkerboard.

The logic that the block implements is a multiple output OR gate with selective input polarities, denoted as a Generalized OR (GOR) gate. Obviously GOR is complete in representing any logic function because of the invertors at the inputs. A GOR block uses the poly-silicon layer and metal 1. Its physical size can be characterized by the number of inputs $n_I(b)$, the number of outputs $n_O(b)$ and the following constants:

W_I	the width of an input line.
W_O	the width of an output line.
W_{ISW}	the minimum width of the input switch and polarity selector.
W_{OBF}	the minimum width of the output buffer and the pre-charging PMOS.

An important feature of the GOR block is that its input lines are fully permutable, as well as its output lines. This provides the freedom that external nets can be routed to the block instead of to the pins, and then the routed wires will determine the pin locations (pin ordering) on the block. All pins are on metal 1.

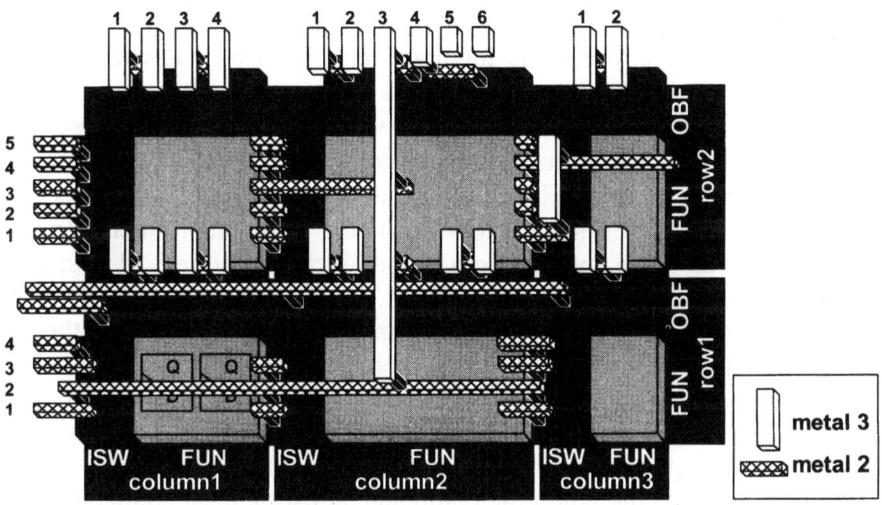

Figure 3.31: The interconnection system of the Checkerboard.

As illustrated in Figure 3.31, the Checkerboard structure is an array ($n_X \times n_Y$) of blocks. The blocks in a column have the same width; and the blocks in the same row have the same height. Note that the column1/row1 block is not a GOR block but a register block (FF). A FF-block has a layout configuration very similar to the GOR-block, except that the OR function part is replaced by registers, pre-charging and input switching circuits are removed and the "start" signal replaced by clock signal. A register occupies more area than an OR; hence, although we still use $n_O(b)$ and $n_I(b)$ to describe the shape of the block, the real effective inputs and outputs are characterized by $n_{FF}(b)$, the number of registers in FF-block(b). Due to the discrete nature of the number of registers and some constraints from the design rules, the relationship between $n_O(b)$, $n_I(b)$ and $n_{FF}(b)$ are normally characterized by a table. For the technology we use in our experiments, the following approximation can be used:

$$n_{FF}(b) = \left\lfloor \frac{n_I(b)}{3.5} \right\rfloor \times \left\lfloor \frac{n_O(b)}{3} \right\rfloor$$

Interconnections between blocks are on metal 2 (horizontal) and 3 (vertical). Metal 2 wires have their routing pitch equal to W_I, such that a horizontal metal 2 wire overlays an input line in the GOR block. However, metal 3 wires have a pitch equal to $2W_O$, meaning that an output line in the GOR block corresponds to two vertical metal 3 wire tracks. Each column can be further partitioned into the input switching column (ISW) and the function column (FUN), and each row into the function row (FUN) and the output buffering row (OBF). The spine topology is adopted to construct the signal nets. The output pin of a net connects to a vertical metal 3 segment, and all input pins of the net connect to the spine via horizontal metal 2 ribs. Signal routing only takes place in the FUN columns and FUN rows. A net can have only one rib in a row, shared by all blocks in the row having input pins on the net. If two gates in the same block have the same input literal (polarity matters), they can share one input pin of the block. In Figure 3.31, a spine net is shown, placed in the 3-rd vertical metal 3 track in FUN-column 2. One rib of the net is in FUN-row 2 at horizontal metal 2 track 3, and the other is in FUN-row 1 at track 2. The second rib connects the input pins of the three blocks in the row. When an input signal is binate in a GOR block (some OR functions in the block need positive inputs while others need negative, as G1, G2 and G3 in Figure 3.32), two input lines are occupied. One receives the external input; while the other, the complement, uses a local connection to get the input. The local connection is routed in the ISW column, as shown in ISW-column3/FUN-row2 in Figure 3.31. There, the fourth input pin takes the external input, and the first pin locally connects to the fourth pin via a metal 3 segment in the ISW column. Since two vertical metal 3 tracks correspond to one output pin, if the two adjacent metal 3 wires (odd/even, if indexing starts from 1) both connect to output pins of the same block, one of them can get a direct connection but the other must detour to reach another available pin. As shown in ISW-column2 /FUN-row2 in Figure 3.31, the third and fourth metal 3 spines both need a connection to an output pin of the underneath block. However one of them, here the fourth one, must use a local metal 2 horizontal segment in OBF to connect to another output pin. The FF blocks have less pins than the overlaying metal 2 and 3 track numbers in both input and output sides; therefore similar local connections are often needed. The ISW columns and OBF rows are also where the "start" signals, clock net and I/O connections are routed. The values of W_{ISW} and W_{OBF}, according to the design technology usually allow at least 5 wiring tracks in them.

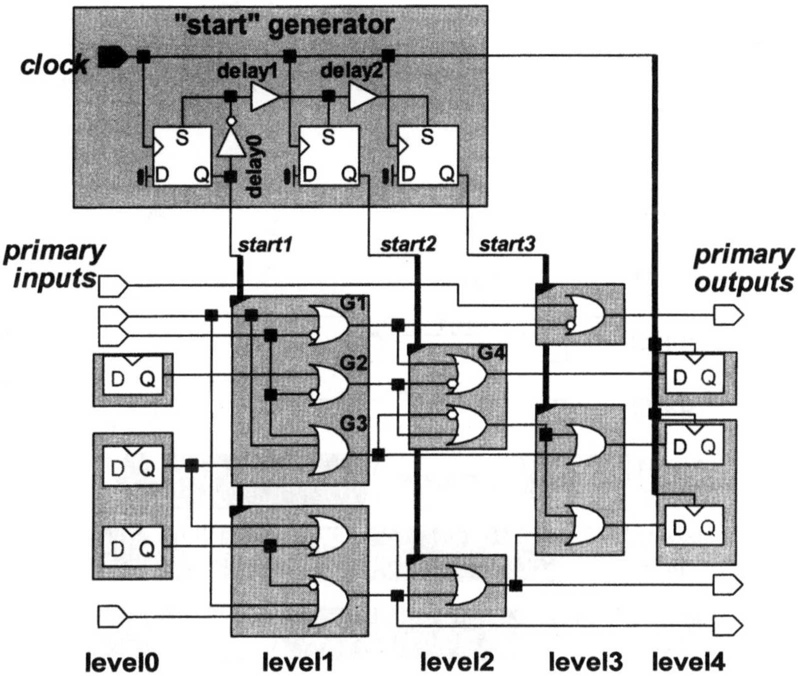

Figure 3.32: The dynamic operation of the Checkerboard.

The Checkerboard implements a sequential circuit (a combinational circuit is just a simplified version with no registers) with a dynamic configuration. Figure 3.32 illustrates the schematic of a typical sequential circuit. The netlist is composed of OR gates, registers and primary inputs and outputs. It is levelized; primary inputs given level 0. Registers are on the highest level, $n_{L\text{-}MAX}$, with only D-inputs visible but they also appear on level 0 with only Q-outputs visible. The primary outputs are placed in level $n_{L\text{-}MAX}$. All the gates are on levels 1 to ($n_{L\text{-}MAX}-1$). The gates on the same level are grouped into clusters and each cluster is implemented with a GOR-block in the CB. Registers are also grouped into FF-blocks. Objects on different levels cannot be placed in the same block. Some gates, such as G4 in the example, may have more than one level that it can be in. Such level flexibility allows the gates to be placed in all the blocks that are "level valid" for them. However, the level flexibilities of different gates may be inter-dependent, because of the fanin-fanout relationships. All the blocks on the same level are controlled by the same "start" signal. The "start" signals are generated in a module called a "start generator". The "start" generator physically occupies one of the blocks in the CB. Delay components, buffers and inverters, create proper timing for the dynamic circuit to work correctly. We assume that the delay components can be continuously sized to get the required delays.

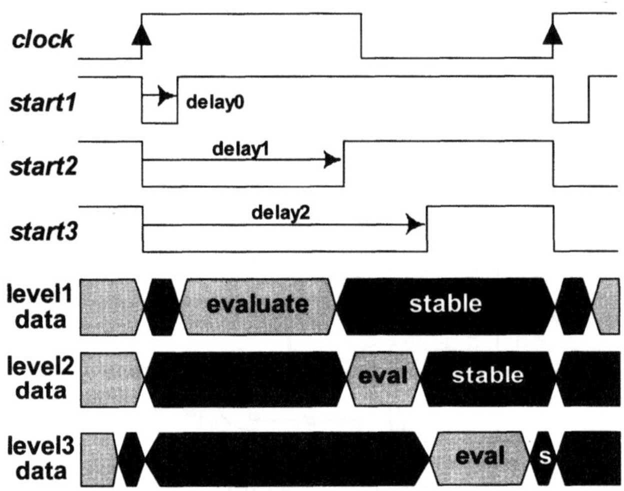

Figure 3.33: The timing of the dynamic operation.

The basic rule of the dynamic circuit is that the evaluation is done level by level in ascending order. A level should not start to evaluate before the data of all lower levels become stable. Figure 3.33 shows the timing of the dynamic Checkerboard. In each clock cycle (rising edge is the reference), each level experiences pre-charging, evaluation and "data stable". The next clock edge should not arrive before the data of level ($n_{L\text{-}MAX}-1$) become stable. During the pre-charging period of level 1, no gate in the circuit is evaluating. Therefore this period of time is "wasted", and it should be kept as small as possible. For pure combinational circuits, there is no clock input but a "start" signal from outside. One more delay stage in the "start" generator is added, simulating the delay of the last level. Its output serves as a "done" signal. Such a configuration can be modified easily to allow the circuits to work also in an asynchronous manner.

The area of a Checkerboard is

$$S_X = \sum_{c=1}^{n_X} \left[n_{FUNC}(c) W_O + \max\left(n_{ISW}(c) Pitch(3), W_{ISW} \right) \right] + \left(n_{LCOL} + n_{RCOL} \right) Pitch(3)$$

$$S_Y = \sum_{r=1}^{n_Y} \left[n_{FUNR}(r) W_I + \max\left(n_{OBF}(r) Pitch(2), W_{OBF} \right) \right] + \left(n_{BROW} + n_{TROW} \right) Pitch(2),$$

$$Area = S_X S_Y$$

in which, $n_{FUNC}(c)$ is the number of GOR-output tracks or half of metal 3 track number in FUN-column c, $n_{FUNR}(r)$ is the number of GOR-input or metal 2 tracks in the FUN-row r, $n_{ISW}(c)$ is the number of metal 3 tracks in ISW-column c, $n_{OBF}(r)$ is the number of metal 2 tracks in OBF-row r, n_{LCOL}, n_{RCOL}, n_{BROW}, n_{TROW} are the numbers of tracks in the four I/O columns/rows respectively, and $Pitch(2)$ and $Pitch(3)$ are the wiring pitches of metal 2 and 3. A block may not have all its inputs and outputs utilized by logic functions; neither are the metal 2 and 3 tracks fully utilized. Variables $n_{FUNC}(c)$, $n_{ISW}(c)$, $n_{FUNR}(r)$ and $n_{OBF}(r)$ are obtained from the packing results of the columns and rows:

$$\begin{aligned}
n_{FUNC}(c) &= \max\left\{ \left\lceil \frac{n_{FUNC-PACK}(c)}{2} \right\rceil, \max_{r=1}^{n_Y} n_O(r,c) \right\} \\
n_{ISW}(c) &= n_{ISW-PACK}(c) \\
n_{FUNR}(r) &= \max\left\{ n_{FUNR-PACK}(r), \max_{c=1}^{n_X} n_I(r,c) \right\} \\
n_{OBF}(r) &= n_{OBF-PACK}(r)
\end{aligned},$$

in which, $n_{FUNC\text{-}PACK}(c)$ and $n_{ISW\text{-}PACK}(c)$ are the numbers of metal 3 tracks in FUN-column and ISW-column c, $n_{FUNR\text{-}PACK}(r)$ and $n_{OBF\text{-}PACK}(r)$ the numbers of metal 2 tracks in FUN-row and OBF-row r, $n_I(r,c)$ and $n_O(r,c)$ are the numbers of input and output lines required by GOR block at row r and column c.

The dynamic configuration requires that the computation is carried out level by level. This implies that the total delay, or the minimum clock cycle, is the sum of the delays of all levels from 1 to ($n_{L\text{-}MAX}-1$) plus the minimum pre-charging time for level 1. The total delay is

$$Delay = d_{PCHG1} + \sum_{l=1}^{n_{L-MAX}-1} d_{LEVEL}(l) = d_{PC1} + \sum_{l=1}^{n_{L-MAX}-1} \max_{Level(g)=l} d_{GATE}(g),$$

in which, d_{PCHG1} is the pre-charging time for level 1, $d_{LEVEL}(l)$ is the delay of level l and $d_{GATE}(g)$ is the delay of gate g. Strictly speaking, $d_{GATE}(g)$ should be measured from the arrival of the "start" signal of the level to the latest arrival time among all its fanouts (before the switches, but after the polarity selectors). Wire effects have to be considered, including the propagation delay from the "start" signal generator to the "start" input pin of the block and the propagation delay from the buffered output to the input pin of the fanout gate. In fact, the "start" signals can leave the generator early to compensate for the propagation delays. Therefore for simplicity, only the wire delays of the regular signal wires are included in our computation. The delay of a gate inside a block uses the linear model given in Section 3.3, that is, the sum of a load independent part and a load dependent part:

$$d_{GATE}(g) = D_G + \delta_{L1} \max_{i \in Input(g)} n_{TRI}(i) + \delta_{L2} n_{TRO}(g) + \delta_{L3} c_{ELOAD}(g) + \max_{fo \in Fanout(g)} \left[d_{WIRE}(fo) \right],$$

in which, D_G is a constant and δ_{L1}, δ_{L2} and δ_{L3} are loading coefficients. In the equation, $n_{TRI}(i)$ is the number of NMOS transistors connected to the input line i, $n_{TRO}(g)$ is the number of input pins of gate g or the number of NMOS transistors connected to the output line, $c_{ELOAD}(g)$ is the total external loading capacitance on the output buffer, and $d_{WIRE}(fo)$ is the propagation delay from the output buffer to the input pin of fanout fo. The pre-charging time for level 1 is computed as

$$d_{PCHG1} = D_{PC} + \delta_{LPC1} \max_{Level(b)=1} \left[n_O(b) + \delta_{LPC2} \max_{Block(g)=b} n_{TRO}(g) \right],$$

in which, D_{PC} is a constant, and δ_{LPC1} and δ_{LPC2} are loading coefficients. Its load is $n_O(b)$ PMOS transistors on the output lines.

3.7.2. The algorithm

The input netlist is technology-independent transformed first. The combinational part is decomposed into a network of GOR gates. The new netlist is passed to the physical design algorithm. The number of blocks in the CB and the number of the blocks on each level are determined. The gates and registers are randomly assigned to their level-valid blocks. The a simulated-annealing improves the initial assignment of levels via moving a gate or register to another level-valid block, swapping two blocks along with their gates and swapping a pair of I/O ports. Based on an assignment of the gates and registers in the blocks and the relative locations of the blocks in the CB, the placement of the gates and registers and their interconnections can be quickly constructed. Area and delay can be then evaluated very precisely.

3.7.2.1. Logic synthesis

Logic synthesis uses *SIS*, an existing synthesis package [SENT92, BRAY90]. After technology independent logic transformation, the levels of the Boolean network are adjusted to make a trade-off between area and delay. Then the network is decomposed into a netlist of OR gates with binate inputs with the *SIS* command "tech_decomp –o". The netlist may contain a few gates with many inputs. Such gates result in wide FUN-rows, which potentially waste area. However, arbitrarily decomposing wide gates into trees of narrower gates may cause the netlist level to increase. Therefore, the netlist is levelized, and each wide gate is examined for flexibility in its levels. If a wide gate has flexibility in its levels, decomposing it has no effect on the netlist level. If the wide gate itself has no level flexibility but some of its inputs come from gates at least two levels lower than the level of the gate, these inputs are extracted to form an OR at one level below. If none of the above two cases occurs, the wide gate are kept as is.

3.7.2.2. Determining the number of blocks in the CB

Let $n_{IP\text{-}MAX}$ be the number of inputs of the widest gate, and assume initially all blocks have $n_{IP\text{-}MAX}$ input pins and $n_{IP\text{-}MAX}$ output pins. Among the $n_X \times n_Y$ blocks, ignoring the "start" generator block, n_{BFF} are register blocks:

$$n_{BFF} = \frac{n_{FF} A_{FF}}{n_{IP-MAX}^2 W_O W_I},$$

in which n_{FF} is the number of registers and A_{FF} is the area of a register. Given *asp*, the desired aspect ratio of the CB, the following equation is obtained:

$$\frac{n_Y \left(n_{IP-MAX} W_I + W_{OBF}\right)}{n_X \left(n_{IP-MAX} W_O + W_{IWS}\right)} = asp\,.$$

Each gate has one output pin, therefore the gate blocks should provide at least n_{GATE} output pins,

$$\left(n_X n_Y - n_{BFF}\right) n_{IP-MAX} = n_{GATE},$$

in which n_{GATE} is the number of gates in the netlist. Let $n_{IP\text{-}TOT}$ be the total number of gate input pins. To make the blocks provide enough input pins,

$$\left(n_X n_Y - n_{BFF}\right) n_{IP-MAX} S = n_{IP-TOT},$$

in which S accounts for the sharing of the input pins by gates in the same block, causing the total number of input pins of the blocks to be smaller than $n_{IP\text{-}TOT}$. With S= 2, we obtain:

$$n_X = \sqrt{\frac{n_{IP-TOT}\left(1+\frac{n_{FF}A_{FF}}{n_{GATE}n_{IP-MAX}W_O W_I}\right)\left(W_I+\frac{W_{OBF}}{n_{IP-MAX}}\right)}{2asp\cdot\left(2\frac{n_{GATE}n_{IP-MAX}W_O}{n_{IP-TOT}}+W_{ISW}\right)}},$$

$$n_Y = \sqrt{\frac{asp\cdot\left(1+\frac{n_{FF}A_{FF}}{n_{GATE}n_{IP-MAX}W_O W_I}\right)\left(2n_{GATE}W_O+\frac{n_{IP-TOT}W_{ISW}}{n_{IP-MAX}}\right)}{2\left(n_{IP-MAX}W_I+W_{OBF}\right)}},$$

and

$$n_{BFF} = \frac{n_{FF}A_{FF}n_{IP-TOT}}{2n_{GATE}n_{IP-MAX}^2 W_O W_I}.$$

The next problem is how to assign these $n_X n_Y$ blocks to all the levels. The netlist has been levelized. We try to determine the number of blocks for level l by the ratio between the number of gates on level l and n_{GATE}. However the levels of some gates are flexible, making the ratio calculation complicated. The most complicated part is that the level flexibilities of different gates might be dependent. Let $l_{MAX}(g)$ and $l_{MIN}(g)$ be the highest and lowest levels of gate g. A triangular distribution is assumed, that is, the probability of taking the center level(s) is the highest and it falls linearly on both sides. Denote the probability of levels $l_{MAX}(g)$ and $l_{MIN}(g)$ by p_{MIN}. Level l, $l_{MIN}(g) \le l \le l_{MAX}(g)$, has a probability of

$$p(l) = \begin{cases} 1.0 & l_R = 1 \\ 0.5 & l_R = 2 \\ \frac{2-(l_R+1)p_{MIN}}{v_R-1} - 4\left|l-l_H\right|\frac{1-l_R p_{MIN}}{(v_R-1)^2} & l_R\text{odd} \\ \frac{2-l_R p_{MIN}}{l_R} - 4\left\lfloor\left|l-l_H\right|\right\rfloor\frac{1-l_R p_{MIN}}{l_R(l_R-2)} & l_R\text{even} \end{cases},$$

in which, $l_R = l_{MAX}(g) - l_{MIN}(g) + 1$ and $l_H = (l_{MAX}(g) + l_{MIN}(g))/2$. This guarantees that the sum of probabilities on all possible levels of a gate is one. We derive, $n^*_{GATE}(l)$, the number of gates on each level l; gates with level flexibilities are counted in probabilities. Then,

$$\frac{n^*_{GATE}(l)}{n_{GATE}}\left(n_X n_Y - n_{BFF} - 1\right)$$

blocks are assigned to level l. Here "-1" accounts for the "start" generator block.

3.7.2.3. Physical design

The physical design is simulated-annealing based, and can be summarized in the following pseudo-code:

```
PhysicalDesign
Simulated annealing framework {
  Randomly do one of {
     Move a gate or register to another level-valid block
     Swap a pair of I/Os.
     Swap a pair of GOR blocks
  }
  Update the input sharing of the affected blocks.
  Routing.
  Evaluate area and delay.
  If (rejected) restore the last placement.
}
```

The input nets of a block are the union of the nets of all the input pins of the gates placed in the block. Note that the number of input pins of a block is no more than the number of its input lines, because some input signals are binate. Input sharing does not occur in register blocks. When a gate moves from one block to another, the two blocks need to update their input pin sharing. Next we discuss the routing.

The key part of the routing is that given the distribution of the blocks of different levels and the assignment of the gates and registers in the blocks, determine the ordering of the input/output lines in the block (i.e. the pin ordering of the blocks) and the interconnections. Fortunately the spine net topology and the column/row regular structure make it possible to do this quickly. The routing interleaves two kinds of operations, planning and packing. The former, corresponding to the global routing in conventional physical design, assigns wire segments like spines and ribs to the columns and rows. Packing corresponds to detailed routing; it arranges the segments in the columns and rows. The following steps are taken during the routing:

- Planning spines/ribs of regular nets in FUN-columns/rows.
- Planning I/O connections.
- Planning "start" signals and clock.
- Packing spines in FUN-columns.
- Packing ribs in FUN-rows.
- Planning local connections.

- Packing in Left/right-I/O columns and ISW-columns.
- Packing in Bottom/top-I/O rows and OBF-rows.

The packing steps do not strictly follow all planning steps because of the dependencies of some information. For instance, only when packing in FUN-rows is done are the locations of the input wires to the GOR blocks known. Then the planning of the local connections between the input wires and input pins can start. The packing algorithm is first described, and the planning procedures follow.

3.7.2.3.1. Packing of segments in a column or row

We adopt the left edge algorithm [SHER93]. This greedy algorithm gives an optimum solution for the packing of segments in a row or column.

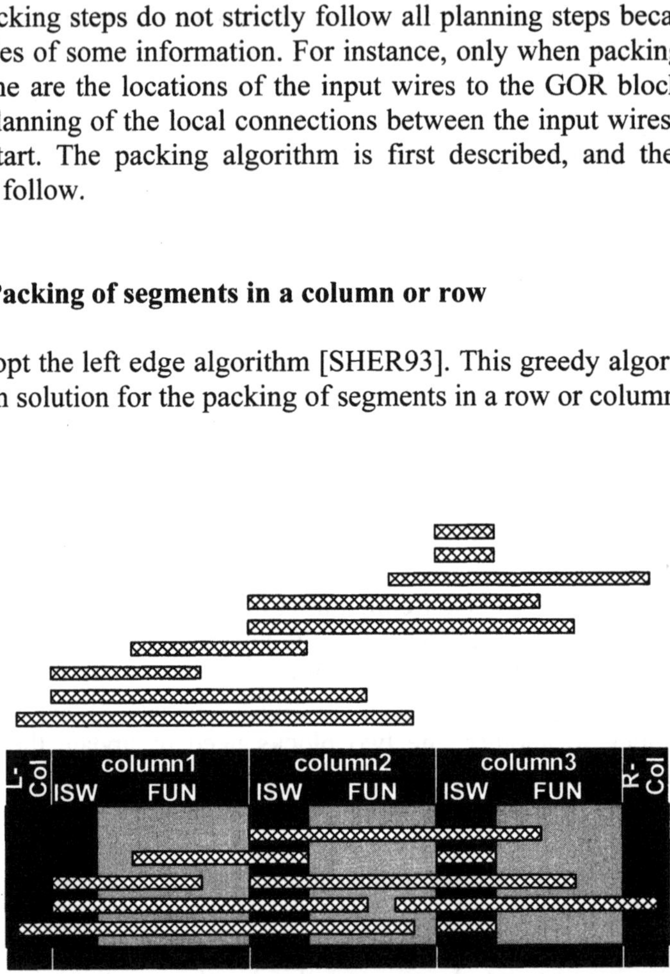

Figure 3.34: Interval packing example.

A packing example of a FUN-row is shown in Figure 3.34. The segments in the row are ordered in ascending order of their left terminals. An unassigned segment that fits in the current track is greedily chosen from the list in the order. If no more tracks can be added to the current track, a new track is introduced. Repeat until all segments are assigned.

3.7.2.3.2. Planning spines/ribs of regular nets in FUN-columns/rows

Given the assignment of the gates and registers, we know, for each net, the FUN-column of the spine and the FUN-row(s) of the rib(s). Note that a net can only have one rib in a FUN-row. The spines inside each FUN-column are collected; so are the ribs in each FUN-row. This corresponds to the global routing in conventional physical design, i.e. a planning of the topologies of the nets on a coarse routing grid. Then each column and row is packed by the "left edge" algorithm, determining the final locations of the spines/ribs in the columns/rows. However, at this moment, a spine does not have exact information about the Y-positions of its bottom and top terminals, the lowest and highest ribs. Only the left or right terminal X-position of a rib is known, i.e. where the input pin is. The other terminal, where the spine is, can be at any X-location inside the FUN-column of the spine. The conflict is resolved by conservatively setting the bottom/top terminals of the spines, which means a spine is assumed to fully span to the top of its highest FUN-row and to the bottom of its lowest FUN-row. This is safe, at the expense of the loss of possible (but small) vertical track sharing. Spine packing precedes the rib packing. So when spine packing is done, all ribs have their left/right terminals known. Finally, for each binate input to a GOR block (both polarities are used in the block), one horizontal segment needs to be inserted in the FUN-row covering its ISW column. These preserve the connecting points for the complemented inputs of the binate signals. In Figure 3.34, the segments spanning the width of ISW-column 3 are for this purpose. The location of connections will be detailed later.

3.7.2.3.3. Planning I/O connections

On the left and right sides of the CB, a left-I/O column (L-Col) and a right-I/O column (R-Col) are built. On the bottom and top sides of the CB, a bottom-I/O row (B-Row) and a top-I/O row (T-Row) are built.

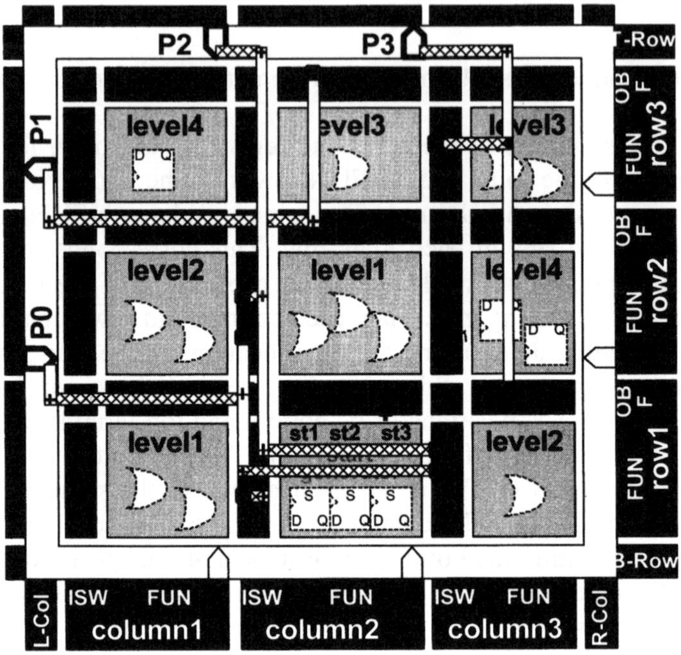

Figure 3.35: The planning of the I/O connections.

The way that I/O connections are made is illustrated in Figure 3.35. *P3*, an output port (with input pin) on the top, places its metal 2 rib in T-Row, and the spine of the net extends to the T-Row to fulfill the connection. *P2*, an input port (with output pin) on the top, places its spine in the nearest ISW-column, ISW-Column 2 in the example. The pin on the port and the spine are connected by a pseudo metal 2 "rib" in T-Row. On the left side of the CB, *P0*, an input port (with output pin) provides the spine of the net. The column the spine is placed in, is determined by taking the average of all the pin columns of the net, assuming the pin on the port itself stays in column 0. To reach the spine, the port first uses a vertical metal 3 segment in L-Col to get to the nearest OBF-row, OBF-Row 1 in the example, and then gets to the spine though a horizontal pseudo rib in the OBF-row. Output port *P1* (with input pin) places its rib in the nearest OBF-row through a metal 3 segment in L-Col. The connections of the I/O ports on the bottom and right sides can be derived similarly. The above procedure puts wire segments into L-Col, R-Col, B-Row, T-Row, ISW columns and OBF rows.

3.7.2.3.4. Planning "start" signals and clock

The clock net routing problem here is simpler than that for standard-cell designs, because in CB, registers are clustered into n_{BFF} FF blocks. Thus, the number of terminal points of the clock is greatly reduced. In addition, the delay adjusting buffers in the blocks can compensate for the skews after the routing of the clock net, if the skew is within a certain range.

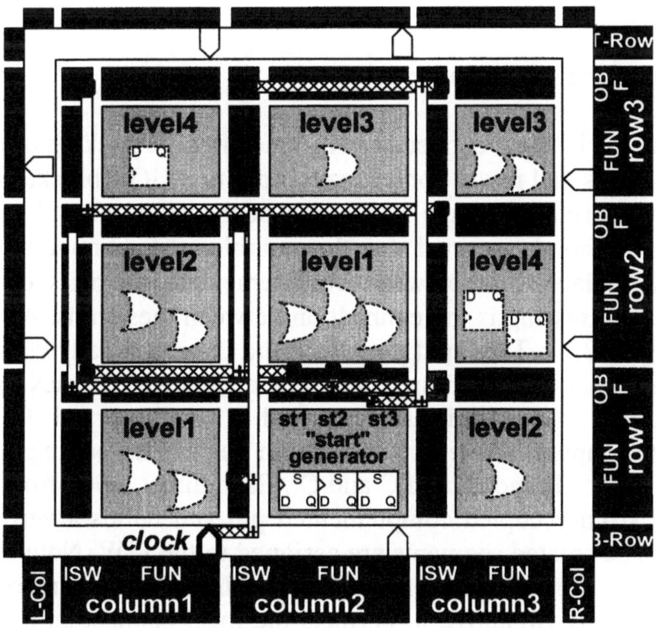

Figure 3.36: The planning of the "start" signals and clock.

Due to the size limit of the CB for efficient implementation (about 10k-gates), wire delays are small although not negligible. Thus, the use of the delay adjusting buffers circumvents the use of a complicated clock routing algorithm. We run a simplified version of Iterated One Steiner Algorithm [KAHN92] on a routing grid formed by the cross-points of ISW columns and OBF rows. A subtle point is that when the clock net is planned, the packing of the ISW and OBF has not started yet. So the distance between the adjacent grids, is not known exactly. Although assuming that the ISW columns have minimum width W_{ISW} and OBF rows have minimum width W_{OBF} is an acceptable approximation, discrepancies may still occur. This means some skew compensation is still necessary, even if an ideal clock planner is used. "Start" signals are "clock" signals generated by the "start" generator, so they are planned as the clock net. The only difference is that the clock starts from an input port, while the "start" signals start from the output pins of the "start" generator.

An example is shown in Figure 3.36. The planning of the "start" signals and clock assigns segments to ISW columns, OBF rows and one of the I/O columns/rows.

3.7.2.3.5. Planning local connections

The planning of local connections is done after the packing of the wires in the FUN-columns and rows. For a binate input to a GOR block, the input pin of the net is determined by the Y-position of the rib as a result of the FUN-row packing. Also determined are the complemented input pin locations. However, the relationship between complemented pin pairs can be freely chosen. What is missing is the connection between these pins. An example is shown in Figure 3.37. The input pins A, B and C have already been connected by ribs on Y-indices 6, 5 and 4; all are binate signals. The packing algorithm also determines that Y-indices 1, 3 and 7 are reserved for the three complemented inputs. However, it does not set up the pairing relationships between A and A', B and B' and C and C'. Each direct input, from bottom up, finds a closest un-occupied complemented pin and builds a vertical metal 3 connection. The created segments are assigned to the ISW. Note that when an input pin is used for complemented input, such as A' at Y-index 1, it is not allowed to run any rib above it, because the local connection already uses a short metal 2 segment at Y-index 1. This also explains the purpose of assigning short metal 2 segments covering the width of ISW to the FUN-row when planning the ribs.

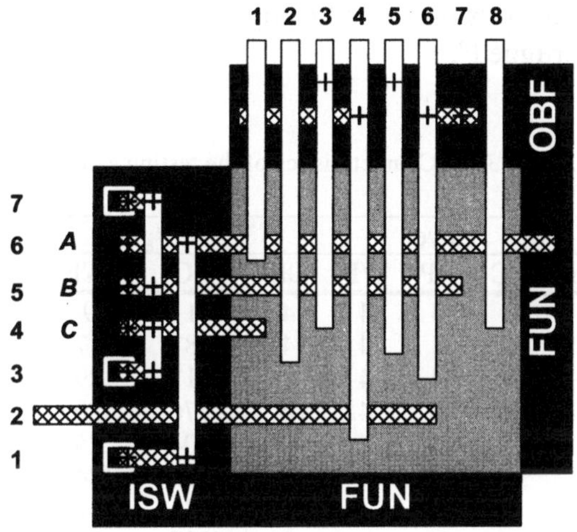

Figure 3.37: Examples of local connections.

The output pins are different. Since two vertical metal 3 tracks correspond to one output line or output pin of the GOR block, only the metal 3 wires at odd X-indices can connect to an output pin directly. Any odd/even pair of spines landing at the same OBF cannot both get output pins. The one with even X-index must find another output pin in the OBF and connect to it through a local horizontal metal 2 segment. The spines landing with even X-index, in ascending order of their X-indices, search from left to right for available output pins. Horizontal segments are created accordingly and assigned to the OBF-row. An example is shown in Figure 3.37. The output pin at X-index 3 has already been connected directly by a spine. The spines at X-indices 4 and 6 both need to connect to output pins. The spine at 4 connects to the output pin at X-index 1; and the spine at 6 connects to the output pin at 7. Note that since the pins at 1 and 7 are reached through local connections on metal 2, a spine can run above it without causing any problem, such as the spine at X-index 1 in the example.

3.7.3. Experimental results

We compared Checkerboard (CB) with standard-cell (SC), River PLA (RP) and Whirlpool PLA (WP) methodologies. A 0.35-μm technology was used. Thirty-one LGSynth91 benchmark examples were tested [LGSY91]. The first twenty-one were made purely combinational, and the rests contain registers. Each circuit was processed using technology independent operations in *SIS* with "script.rugged".

Table 3.15: Characteristics of the testing examples

example	#level					#gate		#PI	#PO	#FF
	SIS	SC	RP	WP	CB	SC	CB			
s208.1(4)	4	9	4	4	4	111	94	10	1	0
s298(4)	4	6	4	4	4	187	172	3	6	0
s382(4)	4	9	4	4	4	317	297	3	6	0
s400(4)	4	10	4	4	4	276	207	3	6	0
s444(4)	4	9	4	4	4	304	258	3	6	0
s526(4)	4	8	4	4	4	302	241	3	6	0
s820(4)	4	11	4	4	4	531	428	18	19	0
apex7(10)	10	19	10		9	522	234	49	37	0
C1355(10)	10	20	10		10	1477	1264	41	32	0
x3(8)	8	12	8		7	1788	707	135	99	0
C2670(18)	18	29	18		15	2213	1518	233	140	0
C5315(16)	16	40	16		16	4027	1934	178	123	0
k2(16)	16	19	16		10	2783	811	45	45	0
k2(8)	8	11	8		7	5675	1059	45	45	0
C5315(16)	16	40	16		16	4027	1934	178	123	0
C5315(12)	12	31	12		12	5247	2958	178	123	0
C3540(18)	18	30	18		18	6363	8466	50	22	0
C7552(18)	18	32	18		15	9180	12766	207	108	0
C7552(12)	12	20	12		12	10894	20034	207	108	0
i10(18)	18	30	18		18	13250	13684	257	224	0
C6288(18)	18	45	28		18	16388	17992	32	32	0
mm4a(18)	18	31			15	118	89	8	5	12
mm4a(10)	10	19			9	127	100	8	5	12
mm4a(8)	8	20			7	175	185	8	5	12
mm9a(12)	12	36			12	443	386	12	9	27
mm9a(10)	10	35			10	487	520	12	9	27
s5378(16)	16	14			11	1415	1362	36	50	164
s5378(12)	12	16			10	1537	1339	36	50	164
dsip(12)	12	11			9	2509	1188	229	198	224
dsip(8)	8	10			6	2519	1390	229	198	224
dsip(6)	6	11			5	2530	1627	229	198	224

The characteristics of the examples tested are listed in Table 3.15. The netlist was then technology mapped with the *SIS* "map" command and then the physical design was performed using ISCPD. RP was only applied to combinational circuits, and WP was only applied to combinational circuits with 4 logic levels due to their structural limitations. For SC and CB implementations, the delay measurement includes wire delays, although neither algorithm used a timing-driven feature in the physical design. In RP and WP, all the interconnections are local, so the wire delays were ignored. All programs were run on a DEC Alpha 8400 5/625 workstation. The area and delay results are given in Table 3.16. In addition, statistics show that the run times of RP, WP and CB are on average 42%, 7% and 106% of the SC run time. In SC and CB, over 95% of the run times were spent on physical design.

Table 3.16: Testing results

	area ($10^3\mu m^2$)				delay (ns)			
example	SC	RP	WP	CB	SC	RP	WP	CB
s208.1(4)	8.2	8.4	2.5	7.3	2.5	1.7	1.9	1.7
s298(4)	12.8	12.1	4.2	13	2.1	2.3	2.6	2.2
s382(4)	22.2	13.6	10.1	18.4	2.7	3.5	3.4	2.5
s400(4)	19	9.2	8.0	18.5	2.6	3.2	3.1	2.4
s444(4)	20.5	12.1	9.0	19.1	2.8	3.2	3.2	2.6
s526(4)	20.8	18.8	9.7	20.2	2.8	3.4	3.3	2.6
s820(4)	30.1	29.9	15.3	27.8	3.9	4.5	4.3	3.9
apex7(10)	40.2	18.8		31.9	6.7	5.4		4.5
C1355(10)	77.2	80.2		83.0	7.4	8.3		6.6
x3(8)	128	150		107	13.6	13.2		10.0
C2670(18)	205	218		194	17.0	17.3		20.4
C5315(16)	324	439		398	26.4	19.0		28.8
k2(16)	188	254		190	15.0	18.2		18.9
k2(8)	486	288		312	19.1	16.1		17.2
C5315(16)	374	439		400	26.1	19.0		22.4
C5315(12)	454	508		471	25.0	19.6		20.8
C3540(18)	430	234		333	18.3	18.4		18.5
C7552(18)	793	820		831	25.4	19.0		21.3
C7552(12)	853	880		1354	21.5	19.5		22.1
i10(18)	962	890		1015	28.3	21.8		26.0
C6288(18)	1586	1759		1703	44.2	53.1		50.0
mm4a(18)	9.3			9.8	25.9			26.7
mm4a(10)	9.3			9.8	8.8			5.5
mm4a(8)	14.0			15.1	9.4			9.6
mm9a(12)	39.9			42.4	30.2			25.9
mm9a(10)	41.3			44.5	27.1			20
s5378(16)	157			172	8.3			12.4
s5378(12)	182			208	9.0			10.1
dsip(12)	307			323	6.8			10.0
dsip(8)	325			351	6.5			7.3
dsip(6)	321			376	7.1			6.9
compare	**1**	0.93	0.42	1.02	**1**	1.02	1.12	0.97

The CB layout of mm9a(10) is illustrated in Figure 3.38.

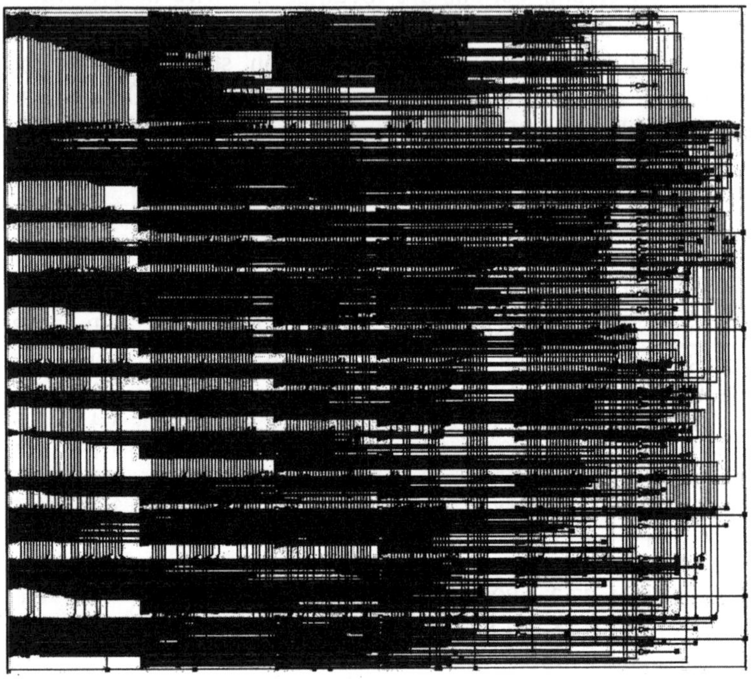

Figure 3.38: The Checkerboard layout of mm9a(10).

An interesting phenomenon observed from Table 3.15 is the change in the level number after technology mapping or decomposition. In the technology independent transformation of *SIS*, a netlist is represented as a network of Sum-Of-Product nodes each of which is equivalent to two logic levels. When decomposing into OR gates, the level number may decrease if some nodes on the longest level path(s) are trivial (only AND or OR function; functions like Z=AB+C are non-trivial). However, the level number after decomposition will never exceed that before decomposition. Usually level number is an important means of controlling the delay of the circuit. Keeping the level number as close (or even smaller) to that of the technology independent result is essential. However the level number of SC circuits after technology mapping is sometimes hard to predict. As discussed earlier, one reason is that some nodes with wide inputs result in trees of gates, because the library cells have very limited numbers of inputs. The other reason is the loading problem. Gates driving too many fanouts may require buffer insertion, which increase the levels. Therefore, the mapping result may deviate from the one predicted by the technology independent transformation.

Comparing the area and delay results in Table 3.16, RP and WP are smaller in area but larger in delay compared to both SC and CB. The area of WP is less than half of the areas of SC and CB; however it can implement only small circuits. In addition, WP only works for 4-level netlists. The CB is on average 2% larger than SC, while it wins 3% in delay.

3.7.4. Summary and discussion

Checkerboard is a regular circuit structure working in a dynamic fashion. The design flow for CB consists of a decomposition into OR-gates with binate inputs and a physical design stage. The structure allows input pins of the same net to be shared by the gates placed in the same block. The spine net topology greatly simplifies the physical design, enabling an integration of the placement and routing. The CB provides a regular structure alternative to the widely used standard-cell structure, with comparable area and delay. The number of levels, a key element in logic synthesis with timing requirement, is well maintained throughout the design flow. Following are some future directions of research.

The CB is suitable for implementing modules with about 10k gates. The limit does not come from the algorithm. Standard-cell designs with three-layer routing system have the same problem. Rent's rule predicts increasing interconnection complexity as gate numbers gets larger [LAND71]. When interconnection dominates the layout resources, the utilization of the "gate" area becomes low. Modern chips contain six or more metal layers. So it is reasonable to build modules with lower metal layers used for internal routing and construct the module-level interconnections on higher metal layers. Chip-level design may involve many modules. The emergence of the CB structure provides an alternative to the SC. As the experimental results show, CB is potentially suitable for circuits with tight timing requirements. A hybrid chip integrating modules built with different structures, as discussed in Chapter 5, might be the best choice that balances overall area, delay and other issues.

The CB can be turned into programmable structures. An extension of CB to a via-programmable implementation has been studied [MO03c]. The sizes of the blocks in a programmable CB are fixed. The wires are also fixed. Only vias are used to determine logic functions and interconnections.

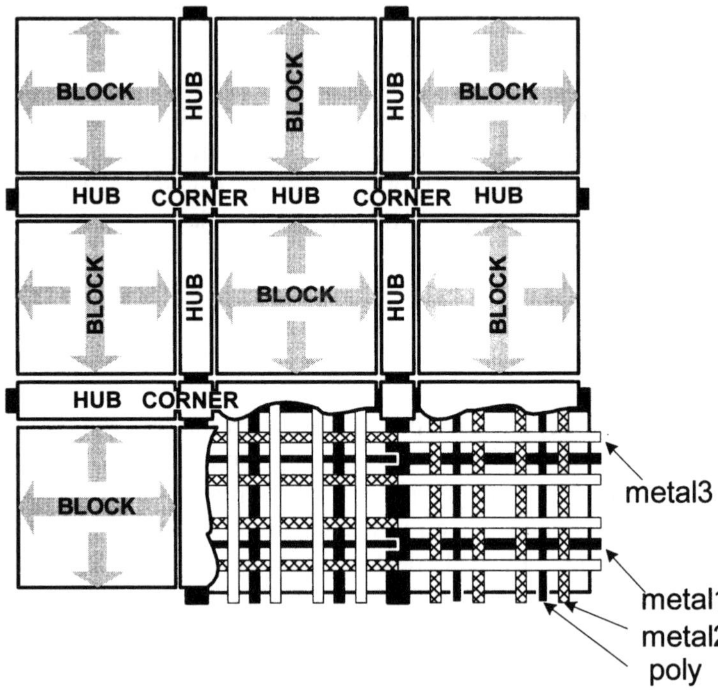

Figure 3.39: Via-Programmable Checkerboard

A diagram of the Via-Programmable Checkerboard structure is shown in Figure 3.39. Global wires, on metal 2 and 3, are segmented, so that a long wire alternates on these two layers when it traverses blocks. Details can be found in [MO03C].

3.8. Comparison of the Structures

This chapter is summarized by comparing the different circuit structures. Table 3.17 compares standard-cell (SC), Network of PLAs (NPLA), Whirlpool PLA (WPLA), River PLA (RPLA) and Checkerboard (CB). The area, delay and run time data are rough values extracted from the experimental results in this chapter. We also examine the possibilities of extending the structures to programmable versions. There are two programmable structures: "field programmable" and "via programmable". Field programmable structures make use of static RAM cells to configure logic functions and interconnections; FPGA is an example. Via-programmable structures use only via layers to configure the logic functions and interconnections, with all other masks pre-designed and fixed. In this sense, normal ICs can be viewed as "mask programmable" structures. GPLA is an example of the via-programmable structure.

Table 3.17: Comparison of the circuit structures

structure		standard-cell	NPLA	WPLA	RPLA	Checkerboard
circuit size		no limit	no limit	1k	10k	no limit
best size (≤3-layer)		around 10k	unknown	1k	10k	around 10k
#metal layers		≥3	≥4	2	2	3
regularity		row	intra -module	full-one -module	full-one -module	full chip
sequential		ok	possible	possible	possible	ok
programmable	field	no	no	no	no	possible
	via	no	no	possible	Glacier PLA	possible
tech. indep. opt		yes	yes	yes	yes	yes
tech. dep. opt		yes	no	no	no	no
placement		integrated	macro-cell	no	integrated	integrated
routing		(ISCPD)	area routing	no	(annealing)	(annealing)
area		1.0	1.2	.42 ~ .63	.93 ~ 1.2	1.0
delay		1.0	1.1	1.1	1.0	.97
run time		1.0	unknown	0.1	0.4	1.1

The regularity of Checkerboard is the highest. WPLA and RPLA have high regularity themselves. But if multiple WPLA and RPLA are integrated, the regularity also depends on the global interconnection structure. This is exactly the situation of the NPLA, for which regularity is limited inside each PLA modules. Standard-cell has the least regularity. We expect the emergence of certain quantitative criteria for the evaluation of manufacturability, at which time the impact of regularity can be assessed more precisely.

None of the PLA-based structures needs technology mapping. WPLA does not even need physical design.

Standard-cell, NPLA and Checkerboard have no limit on circuit size. RPLA has limit on circuit size, because the PLAs in the stack of RPLA cannot be arbitrarily large. WPLA is only suitable for small circuit, for it is a four-level structure. If networks of WPLA and RPLA modules are built just like NPLA, they may be suitable for large circuits. But this may require more metal layers for global interconnection. This implies that when we consider the circuit size that a circuit structure can implement, the number of routing layers is an important factor. The "best size (≤3-layer)" row gives estimated circuit sizes the structures can effectively implement when no more than 3 metal layers are available. With this additional constraint, Rent's rule may indicate that a circuit becomes interconnect dominant as its size grows. We have commented on this in the discussion of the ISCPD algorithm and Checkerboard.

The data show that NPLA is in general inferior to standard-cell. The main reason might be that it requires block-level placement, which results in poor area utilization and cancels the advantage of the compactness of the PLA structures. But as mentioned in Section 3.5, the area might be recovered by filling the white space with logic units, standard-cells for instance. WPLA offers a great opportunity to obtain very compact layout, although the four-level structure limits its applications to small circuits. In addition, its run time is much shorter than those of other structures. RPLA can provide smaller area for some circuits. Also note that it only needs two metal layers for interconnection and the white space in the layout might be further utilized as in the NPLA case. Checkerboard offers similar area but slightly smaller delay, compared to standard-cell.

Comparing the areas using average results from the experiments may lead to incorrect choices of certain structure that seems superior. As already mentioned, the area superiority should not be measured without considering the metal layers used for routing. One may need to consider the circuit size that is being implemented, because a circuit structure might only be good for circuits of certain sizes. Also important is that circuit size relies on the delay requirement. As already shown in many examples, the areas of the same circuit with different delay requirements may differ a lot. Finally whether to use the average statistics to guide the choice of the circuit structure in implementing a specific circuit really depends on the design time allowed. If design time is not a problem, it might be preferable to implement the circuit with all available circuit structures and choose the best one. This may sometimes lead to better result, if the circuit happens to not follow the average statistics.

In some circumstances, design time matters. For instance, if the implementation of a circuit module is part of a design flow and this step is called many times due to iterations, we may not afford to test every time which structure is best. Then we may have to limit in advance the choice to one or a few structures. Taking into account various issues discussed above, standard-cell and Checkerboard might be the two structures that are worth allowing a choice in a design flow. In the design flow we propose in Chapter 5, these two structures are adopted because of their good area/delay quality and flexibility in implementing circuits of difference sizes. Other structures may find specific applications in which they outperform standard-cell and Checkerboard. For example, WPLA can be used for the implementation of a decoder, replacing the common single-PLA approach.

Chapter 4

BLOCK-LEVEL PLACEMENT AND ROUTING

4.1. Overview

Block-level placement and routing is the design stage, in which hard modules have their locations and orientations determined and their interconnections constructed. A hard module can belong to one of following categories:

(1) Parameterized modules made by certain layout compiler, including RAM and ROM.

(2) Intellectual Property (IP) blocks such as embedded processors.

(3) Modules generated with circuit structures such as standard-cell, River-PLA (RPLA), Whirlpool-PLA (WPLA) and Checkerboard (CB).

Block-level placement and routing normally handles the top-level layout. If the netlist contains only hard modules like the IP blocks, block-level placement and routing is applied directly. If standard-cells are contained in the netlist, they can be grouped into modules (soft modules before physical design); each such module is synthesized and physical-designed to become a hard module. If standard-cells join the top level physical design together with hard modules, we call this a mixed block-cell placement and routing.

Block-level placement and routing is adopted in our Module-Based design flow (Chapter 5). Moreover, we integrate the placement and routing into a single algorithm called Fishbone. The routing uses a two-layer spine topology. The pin locations are configurable and restricted to certain routing grids in order to ensure full routability and precise predictability. With this scheme, exact net topologies are determined by pin positions only; hence during block placement, net parameters such as wire delay can be derived directly. The construction of Fishbone nets is fast, enabling the integration of block placement and routing; there is no separate routing stage. A simultaneous buffer insertion algorithm is integrated into the Fishbone scheme. Experimental results show that the Fishbone scheme with buffering produces layout with area and performance comparable to Steiner Tree based algorithm; while its run time is much shorter. The Fishbone scheme is presented in Section 4.2 and Fishbone with buffer insertion in Section 4.3.

4.2. Fishbone Scheme

As System-On-a-Chip (SOC) becomes more and more popular, block-level placement and routing will play an increasingly important role in the design flow. This also facilitates hierarchical physical design, which narrows the gap between the capacity of existing algorithms and growing design size. In traditional design flow, placement and routing are separate stages. The goals of placement include minimizing area, wire delays, and maximizing routability. However, except for layout area, wire delays and routability are based on estimations made during placement. The accuracy of the estimation is a key factor in obtaining good final results; bad estimation may lead to iterations of the two stages. The separation of these stages allows each stage to focus on one or a few targets trying for the best; such a sequential procedure saves run time. The cost is possible non-convergence due to wrong early estimation. The problem of block-level routing is quite different from gate-level routing [YOSH01]. Usually the wiring density at the block level is not as high as the gate level. However routability problems do occur, even if the total routing density is low. One major cause is local congestion in the pin regions. The pins of the blocks are often placed on the block boundaries. When two blocks are placed side by side, the pins on the abutting edges are very close, causing routing congestion. Using more routing layers or leaving extra space between blocks are possible cures; however both are costly and it is hard to determine, especially at the placement stage, how much such extra resource is sufficient.

Estimation of wiring effects during placement is usually based on certain net models. Various models are available [KAHN01, BODA01, MO00, SAIT99], among which the Minimum Steiner Tree (MST) and Half Perimeter (HP) models are popular. Since block-level routers commonly use the MST model, it is preferable to use the same model for estimation in placement. Unfortunately, its run-time cost is too high. Also due to congestion, nets planned individually using MST may require detours, causing deviation from the estimated wire topologies. Other estimation models, such as HP, have shorter run times, but are less accurate. Actually HP is rather a model of wire length than net topology. It has been tightly associated with algorithms targeting minimization of wire length but is less appropriate in expressing wire delay. In timing-driven design flows, a widely adopted method is to define and use net criticality, the priority of a net being optimized for delay in the placement and routing. However it is really hard to know which nets are critical without physical information. Criticalities of the nets may change during the placement. Only when a placement is given and a net model is assumed, can we tell which nets are more critical than others.

We propose a fixed routing topology which is totally determined by the pin positions. Thus whenever a block placement is given, the routing of all the nets is determined precisely; block placement and routing become an integrated task, and no estimation is necessary. Neither is estimation of net criticality required. The fixed net topology we use is called Fishbone; it is simply a spine with the output (source) pin of the net on a vertical trunk (e.g. on metal 4) and all input (sink) pins connected to the trunk by horizontal branches (e.g. on metal 3). It might seem naive to use a topology obviously inferior to the MST since previous research has shown that spine structures are not good net models for approximating Rectilinear Steiner Trees in terms of the "fidelity" of the estimation [GANL97]. Spine topologies are only found in clock and power/ground routing. In general, Fishbone produces longer wire lengths or delays than MST. However, this is true only when the pin locations are fixed. A good block placement found with Fishbone topology should have located the pins such that the associated Fishbone nets have wire lengths or delays comparable to those of the same placement but measured with the MST model. The goal of the Fishbone block-level placement and routing scheme is to find such placements. In other words, if the output of the Fishbone placer is post-processed by a MST router, the improvement will be small. An attractive point of the Fishbone is that the wire topology and thus the wire lengths and delays are totally determined by the placement. Therefore a big difference between a Fishbone-based placer and other block-level placers is that the Fishbone scheme can use precise wire lengths or delays in the cost function of the placement algorithm. The routability of the Fishbone scheme can be tested quickly. This is enabled by requiring, in the Fishbone scheme, the pin locations on the blocks and the block locations on the layout to be constrained to certain grids. Pin locations on the blocks are made configurable by using a wide base pin (e.g. on metal 2) and using the concept of a configurable virtual pin. The current version of the Fishbone scheme uses two metal layers for routing. Each layer only has one routing direction.

The advantages of the Fishbone scheme include better predictability for wire delay, better manufacturability, and flexibility in incorporating it with other physical design algorithms:

(1) The use of a known net topology during the block placement eliminates inaccurate estimation for wire delays. In the Fishbone scheme, the construction of the wiring of the nets and the delay measurement are

fast, allowing its integration into simulated-annealing based block placement.

(2) Manufacturability is becoming more important, as the feature size keeps shrinking [CONG01a]. So far the traditional Computer Aided Design (CAD) for Very Large Scale Integrated Circuits (VLSI) is almost isolated from the backend mask engineering techniques, due to the lack of available models formulating the manufacturing behaviors that can be used by the CAD algorithms. Regular layout patterns are favorable for Optical Proximity Correction (OPC), a major backend technique to enhance the lithography quality [PLUM00], because the number of the unique patterns is reduced [LIEB03]. For a routing algorithm in particular, the requirement is that the wires produced contain less segments and vias and have simpler shapes. Wire length, a common optimization target in the routing, is not much related to this requirement. The Fishbone topology simplifies net shapes and reduces segment numbers.

(3) The Fishbone scheme can be viewed as a framework for block-level physical design. Other optimization algorithms like buffer insertion can be incorporated into the framework. These can benefit from the known net topologies and accurate wire delays at each placement step. But since the algorithm is called in the inner loop of the simulated-annealing, their time complexities need to be controlled.

The disadvantage of Fishbone might be its inferiority in wire length or delay compared with MST. However the problem at hand is not to route for a fixed set of pins. The blocks, carrying the pins, are movable. Hence a more precise comparison of the final area, delay and run time should be made between a Fishbone flow and a fully MST-based placement and routing flow.

The rest of this section is organized as follows. A quick overview of the Fishbone scheme is given in 4.2.1. The basics of the Fishbone scheme are detailed in 4.2.2. In 4.2.3, the interval packing algorithm is discussed, and the I/O connections are discussed in 4.2.4. Section 4.2.5 gives the physical design flow using the Fishbone scheme. Section 4.2.6 gives experimental results, and 4.2.7 summarizes.

4.2.1. An overview of the Fishbone scheme

To show how the Fishbone scheme works without going into detail, we give an example in Figure 4.1. The real pins of a block, called the base pins, are on layer m_B. The base pins are stripes spanning a certain range. The rest area on layer m_B of the block is the obstruction. The routing takes place on layer m_{B+1} and m_{B+2}. The so-called virtual pins, which are points, are what the router can see. A base pin corresponds to a virtual pin, and the location of a virtual pin is slightly flexible because the base pin is a stripe. A net is a set of pins that are electrically connected, including one output pin (driver) and one or more input pins (loads). A spine topology, as described in Section 2.2.4.3, is used. The output pin connects to a vertical trunk, and the input pins connect to the trunk via horizontal branches. Owing to the base-virtual pin pair, the trunk and branches can slide in a small range. Given a placement of the blocks, the base pin locations are known. The trunks are assigned to columns and branches to rows, based on their base pin locations. A simple interval packing algorithm arranges the trunks in each column and branches in each row. The arrangement finally determines the exact points that the virtual pins connect to the base pins.

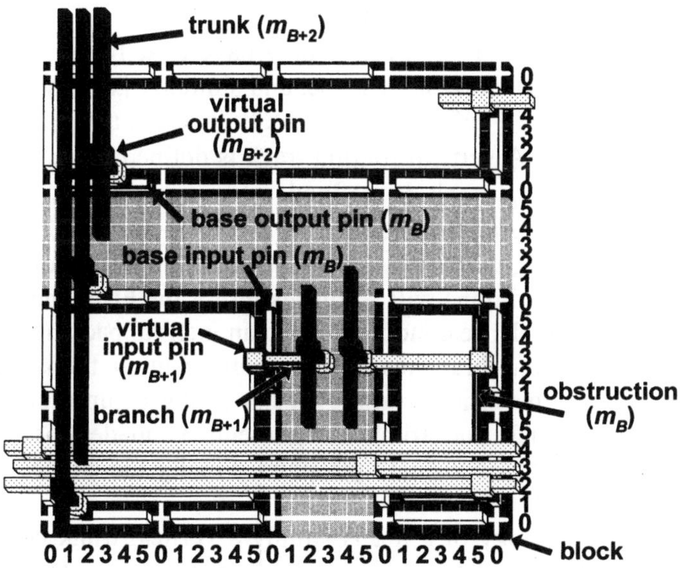

Figure 4.1: The base/virtual pin pair of the Fishbone scheme.

4.2.2. The basics of the Fishbone routing

We assume that the layout uses metal layers up to m_B for internal routing of a block, and that the pins of the blocks and the I/O ports are on m_B. The layers m_{B+1} and m_{B+2} are used for global (Fishbone) routing. It is reasonable to assume that there is no obstruction in the global routing layers. Each such layer uses a uniform routing pitch and a fixed wiring direction (m_{B+1} horizontal and m_{B+2} vertical). The pitches of m_{B+1} and m_{B+2} may be different, but for simplicity we assume that the same pitch is used. The routing grids are given cyclic indices labeled 0,1,2,..,$GR-1$ in both the X and Y directions, where GR is the radix of the grid index. The notation $[x, y, z]$ defines the coordinate of a point. When the grid indices are important, we use the notation $[x^{GX}, y^{GY}, z]$ and the superscripts GX and GY ($0 \le GX, GY \le GR-1$) denote the grid indices. The relation between coordinates and indices can be represented as:

$$GX = \left(\frac{x}{pitch(m_{B+2})} \right) \bmod GR$$
$$GY = \left(\frac{y}{pitch(m_{B+1})} \right) \bmod GR$$

Two coordinates may differ even if they have exactly the same GX and GY. $[x_1 \sim x_2, y, z]$ defines a horizontal stripe, and similarly $[x, y_1 \sim y_2, z]$ defines a vertical stripe. $[x, y, z_1 \sim z_2]$ defines a via between layers z_1 and z_2, if $z_2 = z_1 + 1$, or stacked vias otherwise. A vertical routing track is defined as $[x, -\infty \sim \infty, m_{B+2}]$, and a horizontal routing track is defined as $[-\infty \sim \infty, y, m_{B+1}]$. A column channel (*or column for short*) is defined as $[x^{1 \sim GR-1}, -\infty \sim \infty, m_{B+2}]$ and a row channel (*or row*) $[-\infty \sim \infty, y^{1 \sim GR-1}, m_{B+1}]$, both containing $GR-1$ routing tracks.

Two routability conditions will be discussed for the Fishbone scheme: the basic condition and the dense condition. Then an interval packing algorithm is discussed, which assigns branches to horizontal routing tracks in the row channels and trunks to vertical routing tracks in the column channels. The flexibility of the branch and trunk assignments is made possible by the configurability of the "base/virtual" pin pairs. Finally the connections to the I/O-ports are discussed.

The Fishbone topology of a net is simply a vertical trunk connecting an output (source) pin of the net and horizontal branch(es) connecting this trunk to the input (sink) pin(s). The following definitions describe the base/virtual pin structure, a key element in the Fishbone scheme.

Definition 1 — A ***base input pin*** is a vertical stripe $[x^0, y^{1 \sim GR-1}, m_B]$ with zero X-index. A ***base output pin*** is a horizontal stripe $[x^{1 \sim GR-1}, y^0, m_B]$ with zero Y-index. Base pins are part of the blocks and cannot be moved or modified.

Definition 2 — A ***virtual input pin***, corresponding to a base input pin defined above, is a point $[x^0, y^{GIY}, m_{B+1}]$, connecting through a via down to the base input pin, where G_{IY} is the Y-index of the virtual input pin, ranging from 1 to $GR-1$. A ***virtual output pin*** is a point $[x^{GOX}, y^0, m_{B+2}]$, connecting through two stacked vias down to the base output pin, where G_{OX} is the X-index of the virtual output pin, ranging from 1 to $GR-1$. The pair of base and virtual pin provides configurability for the connecting point to the net.

These definitions are illustrated in Figure 4.2. In the figure, $GR=6$, $G_{IY}(A)=3$ and $G_{OX}(V)=5$. The virtual input pin A, located at $[x^0, y^3, m_{B+1}]$, is made with a via connecting down to its base pin $[x^0, y^{1\sim5}, m_B]$. The virtual output pin V, located at $[x^5, y^0, m_{B+2}]$, is made with two stacked vias connecting down to its base pin $[x^{1\sim5}, y^0, m_B]$. The Fishbone routing deals with virtual pins only. The base-virtual pin mechanism provides flexibility in locating pin position in a small range.

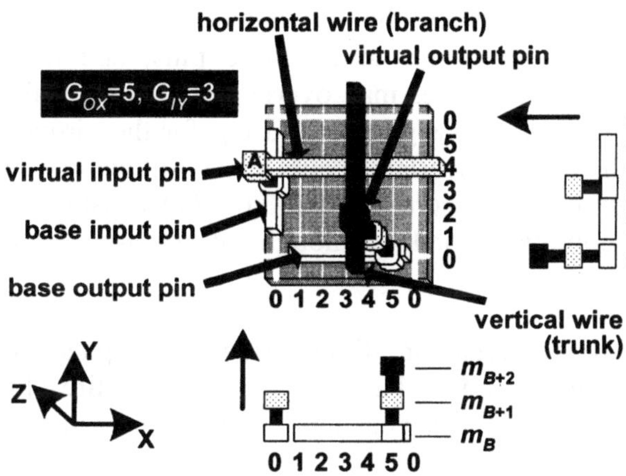

Figure 4.2: The base/virtual pin pair of the Fishbone scheme.

The blocks are designed such that a commonly agreed GR is used (the selection of GR will be discussed later). All blocks used in the same design must use the same GR (we will discuss in Section 4.2.7 how to relax this constraint when integrating blocks designed without a common GR). For simplicity, the block size is assumed to be an integer number of column channels and row channels, given a GR. This extra requirement causes negligible increase in the size of the block since GR is small and the numbers of columns and rows are large. The Fishbone placer will always place the blocks such that their corners are at an X/Y-index of 0/0. Thus a location inside any block can be easily positioned by the global X/Y-coordinates and -indices. The following definition states how a Fishbone net is constructed with the output pin labeled 0 and the input pins labeled 1 to *#IP*. When routing is discussed, we simply use "pin" to refer to "virtual pin".

Definition 3 — Suppose that a net contains the output pin $[x_0^{GOX(0)}, y_0^{0}, m_{B+2}]$ and input pins $[x_j^{0}, y_j^{GIY(j)}, m_{B+1}]$. The ***vertical trunk*** of the net is a stripe $[x_0, \min(y_j)\sim\max(y_j), m_{B+2}]$ for j=0,1,..,*#IP*. An input pin is connected to the trunk through a ***horizontal branch*** $[\min(x_0, x_j)\sim\max(x_0, x_j), y_j, m_{B+1}]$ and a via at $[x_0, y_j, m_{B+1}\sim m_{B+2}]$.

Since $G_{IY}\neq 0$ for any virtual input pin, and all virtual output pins have their vias placed at a 0 Y-index by definition, a horizontal branch will never intersect a via of any virtual output pin. The vertical trunk is on layer m_{B+2}; thus it will never intersect any virtual input pin or horizontal branch on m_{B+1}. Thus the virtual input pins and the horizontal branches will not interfere with virtual output pins, their vias to the base pins, and the vertical trunks. However, branches themselves may overlap; similarly for the trunks. Thus a problem arises on how to assign G_{OX} and G_{IY} for the virtual output pins and input pins such that no overlap occurs. The following is a sufficient but conservative condition for zero overlap, i.e. 100% Fishbone routability. We discuss this condition first, although it over-constrains the design significantly.

Condition 1: For a given *GR*, all the virtual output pins in the same column (recall that columns are global, i.e. across the entire chip) are assigned different G_{OX} and all the virtual input pins in the same row are assigned different G_{IY}.

Proposition 1: Basic Fishbone — A placement is 100% routable if *Condition 1* is satisfied.

This condition over-constrains the routing and possibly wastes routing resources. For instance, each m_{B+2} vertical track on the chip is either fully empty or only occupied by one vertical trunk. An example satisfying the basic Fishbone condition is illustrated in Figure 4.3. In the example, the input pins of a block are placed on the left and right sides of the block, and the output pins are placed on the top and bottom sides of each block (this is not a constraint; pins can be placed anywhere within the block). *Condition 1* also implies that no two input pins of the same block can be placed in the same row, and no two output pins of the same block can be placed in the same column. Thus the numbers of input and output pins that can be accommodated in a single block are equal to the numbers of rows and columns occupied by the height and width of the block. Although small *GR* allows more pins per block, it increases the possibility that the number of pins of different blocks appearing in the same row or column of the layout exceeds *GR*−1 (recall that to satisfy *Condition 1*, a row or column can maximally accommodate *GR*−1 pins).

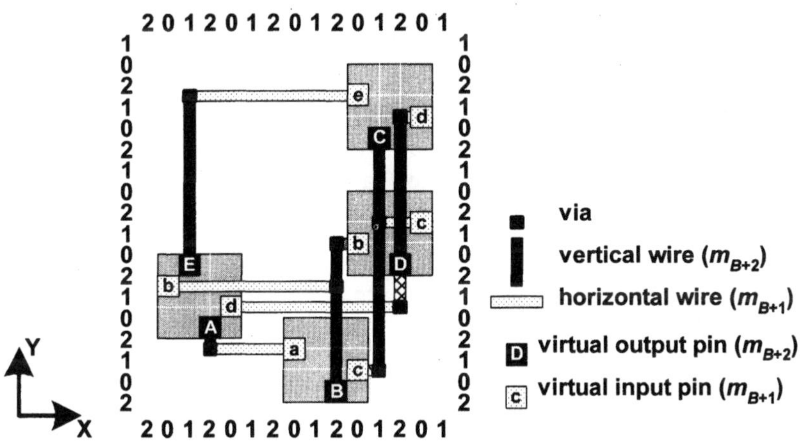

Figure 4.3: Basic Fishbone scheme (*GR*=3) connecting 4 blocks. Only virtual pins are shown; base pins are omitted.

A valid (but bad) placement exists even if *GR*=2, e.g. all the blocks, using G_{IY}=1 and G_{OX}=1 for any input and output pins respectively, are placed in a 45° line such that any column or row is occupied by only one block. In general, it gets harder to find a good placement in terms of area and wire length/delay as *GR* becomes smaller, while using a large *GR* allows better placement, but either limits the number of pins of the blocks, or the block size needs to be expanded just to accommodate all its pins. Thus the choice of *GR* can significantly influence the quality of the result. Also note that *GR* should be defined early in the flow so that it can be used to determine block size and base pin locations and sizes. A reasonable *GR* can be chosen as

$$GR=\lceil \#B^{1/2} \rceil+\rho,$$

where #*B* is the total number of the blocks to be placed with the Fishbone scheme and ρ is a small integer.

An observation is that there is no vertical trunk at $[x^0$, any~any, $m_{B+2}]$, which means G_{OX} of a virtual output pin can safely be 0. In addition, X and Y directions (layer m_{B+1} and m_{B+2} respectively) can use different *GR*'s. However, for simplicity and symmetry, the requirement of G_{IY},G_{OX}=1~*GR* is used. The reason that virtual output pins and trunks are put on a higher metal layer is that the number of output pins in a design is not greater and usually much less than the number of input pins. Such a layer assignment will use less vias than the opposite assignment.

The constraint that no two vertical trunks are placed at the same X-coordinate and no two horizontal branches are placed at the same Y-coordinate can be relaxed; this results in the so-called **dense** Fishbone scheme:

Condition 2: All the vertical trunks with the same X-coordinate (same G_{OX} in the same column channel) are non-overlapping, and all the horizontal branches with the same Y-coordinate (same G_{IY} in the same row channel) are non-overlapping.

Proposition 2: Dense Fishbone — A placement is 100% routable if *Condition 2* is satisfied.

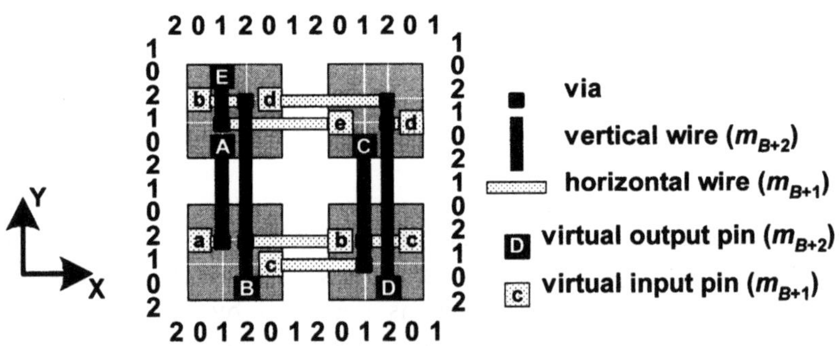

Figure 4.4: Dense Fishbone scheme (*GR*=3) connecting 4 blocks. Only virtual pins are shown; base pins are omitted.

The use of the dense scheme increases the routing resource utilization, while maintaining the Fishbone topology. It is similar to channel routing, in which more than one segment may share the same routing track. Also, more pins can be accommodated in each block by using a smaller *GR*. The example in Figure 4.3 has been re-drawn in Figure 4.4 using the dense Fishbone scheme. Note that now the ribs of a, b and c share the second G_{IY}=2 horizontal track from the bottom. We will discuss in the next sub-section how the dense Fishbone condition can be insured during placement by using an interval packing algorithm. The dense Fishbone scheme is the one adopted in our algorithm on which we report experimental results.

4.2.3. Interval packing for branches and trunks

The arrangement of the branches and trunks determines the Y/X-coordinates of the virtual input/output pins. Since a virtual input/output pin only has the freedom in choosing its Y/X-coordinate within a row/column, the problem is equivalent to determining G_{IY}/G_{OX}. With the *basic* Fishbone condition, the arrangement is trivial; whether the condition can be satisfied is just a matter of counting the number of input/output pins in the same row/column.

For the *dense* Fishbone condition, we determine G_{IY} for input pins first and then determine G_{OX} for the output pins. First consider the G_{IY} determination for the virtual input pins. Given a placement of the blocks, the input pins are already assigned to row channels. The branch of input pin P_i is a horizontal stripe $[x_L(P_i)\sim x_R(P_i), y^{GIY(Pi)}, m_{B+1}]$, in which $G_{IY}(P_i)$ is the Y-index of the virtual input pin, $x_L(P_i)$ and $x_R(P_i)$ are X-coordinates of the left and right terminals of the branch. Either $x_L(P_i)$ or $x_R(P_i)$ is the X-coordinate of the trunk of the net, while the other is the X-coordinate of the input pin (0 X-index). Since the G_{OX} determination for the virtual output pins is done after the determination of G_{IY} of the virtual input pins, the exact G_{OX} or the X-coordinate of the trunk is not known. The range of the X-coordinate of the trunk is $x^1(P_0)\sim x^{GR-1}(P_0)$. Hence, if the left terminal of the branch connects the trunk, let $x_L(P_i)=x^1(P_0)$; and if the right terminal of the branch connects the trunk, let $x_R(P_i)=x^{GR-1}(P_0)$. An interval packing algorithm is employed to arrange all branches in the same row.

The interval packing algorithm gives the minimum number of tracks needed to accommodate all the intervals [SHER93, ZHAN00, HASH71, BURS86]. There are several implementations of the interval packing algorithm. All start with a sorting of the segments in terms of their left X-coordinates, and differ in the subsequent track filling part. The sorting has a complexity of $(n_w\log(n_w))$, in which n_w is the number of segments. The original "left-edge" algorithm greedily fills up a track with un-used segments in the ordered sequence [HASH71]. A new track is introduced when the current track can accept no more segments. Here $GR-1$ tracks are available. The complexity of this track filling procedure is $O(n_w\log(n_w))$, in which n_w is the number of segments to be arranged. Another implementation, instead of filling up tracks one by one, places segments one by one, and each segment is placed in the lowest available track [BURS86]. The complexity is $O(n_w\log(n_d))$, where n_d is the number of tracks. The two algorithms are virtually equivalent, and n_d is in the worst case equal to n_w. In the Fishbone scheme, the "channel" height is fixed, i.e., $GR-1$ tracks. It is not necessary to continue the interval packing algorithm if the number of tracks needed already exceeds $GR-1$. We use the second packing algorithm.

With respect to the way the overflowing segments are treated, it can be further classified into three versions. In the first version, as a segment needs to be placed in track GR, the packing stops and all the remaining segments, including this overflowing one, are treated as overflowing segments. The second way is that the packing still continues, even if some segment overflows. All the overflowing segments are excluded from packing and they are counted as violations. The second approach gives a more precise value of the overflowing segments (not tracks), although the first is faster in some cases. Both approaches have a complexity of $O(n_w)$ in filling tracks, and each may overestimate the violations. The third way is to just do a full packing as if there is no limit on the number of the available tracks. It is the slowest method but gives the exact number of the overflowing tracks. Note that in the Fishbone scheme, the number of segments packed in a channel is usually limited. Despite the fact that the overall complexity is theoretically dominated by the sorting, reducing the time spent on the track filling part is very important. On the other hand, the number of violations does not need to be exact (overflowing tracks) but only a reasonable approximation. The second method is adopted, balancing the run time and the accuracy of violation counts.

The track filling part of the algorithm is summarized as follows. Let n_V denote the number of violations which is set to zero at the beginning. For track d ($1 \le d \le GR-1$), a variable $RT(d)$ is built to record the rightmost coordinate that is covered by previous segments in track d. All $RT(d)$'s are initialized to zero. Variable n_d represents the current number of tracks used and is initialized to 1. Pick up a segment from the sorted list in the ordered sequence. Check from track 1 to n_d to see if any track can accommodate the segment without overlap, by comparing the left of the segment and $RT(d)$. If found, fill the segment in that track and update the corresponding $RT(d)$. If none of the n_d tracks has vacancy for the segment, track n_d+1 is introduced and the segment is placed. If n_d has already reached $GR-1$, just assign the segment to track GR (a "null" track) and increment n_V by one. The reason for assigning overflowing segments to the "null" track is that the subsequent interval packing for the vertical trunks in the columns requires the ranges of the Y-coordinates of the trunks, which are determined by the Y-coordinates of the branches.

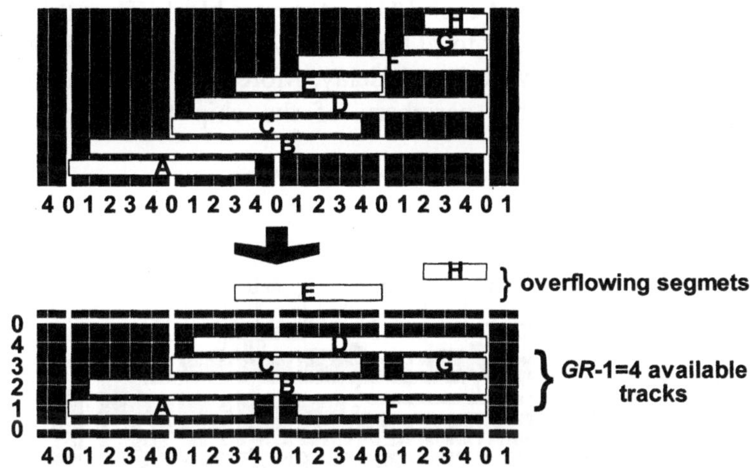

Figure 4.5: An interval packing example.

The algorithm is illustrated with the example shown in Figure 4.5. Segment E and H overflow, so there are two violations.

After the branches are arranged in rows and the Y-coordinates of the virtual input pins are determined, the placement of trunks or the determination of X-coordinates of the virtual output pins begins; this also using interval packing. However since the branches are already determined (possibly some branches overflow, but they still have Y-coordinates, i.e., the "null" tracks with indexes of *GR*), a trunk has its Y range known exactly.

The above algorithm returns an arrangement of branches and trunks for the given block placement and can report the number of violations (overflowing segments). The goal of Fishbone block placement, besides seeking a minimal combination of area and wire delay, is to find a solution with no violations. Note that since this routing scheme is very fast, it can be used during placement to provide integrated placement and routing algorithms.

4.2.4. I/O-pins

A remaining problem is how to handle the pins on the I/O ports of the chip. Assume that an I/O port carries exactly one pin, called an I/O-pin. Also assume that the I/O ports are evenly spread along the chip boundary. An obvious choice is to place all input ports (carrying output I/O-pins) on the top and bottom boundaries and all output ports (carrying input I/O-pins) on the left and right boundaries, as if each I/O port were a block. Then the Fishbone scheme can be applied directly to I/O-pins. However this simple approach has an adverse impact on the wire length/delay. Moreover, a designer may give the I/O sequence along the periphery of the chip for reasons such as packaging constraints.

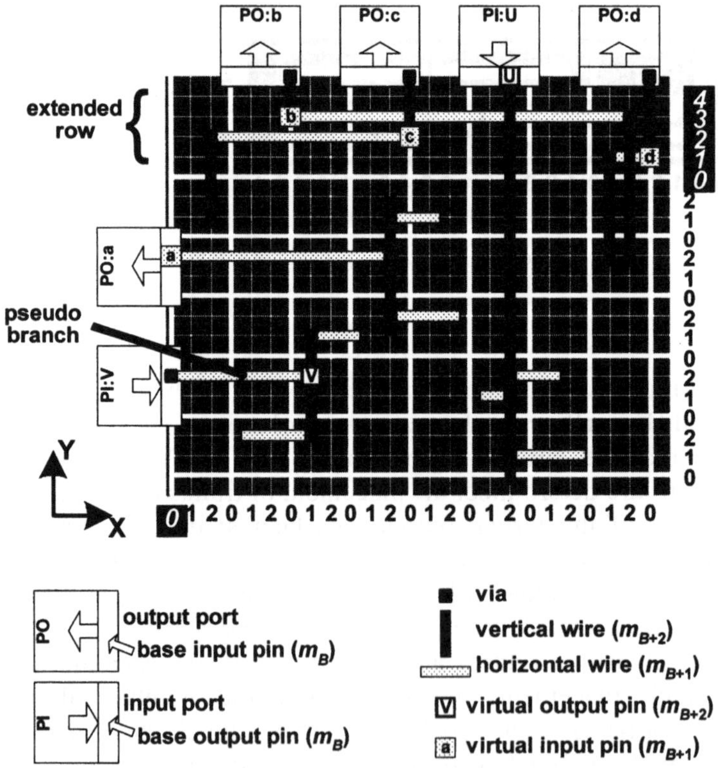

Figure 4.6: The I/O-pins and their connections.

To allow I/O ports to be placed on any side of the boundary, we need to consider two special cases. One is an input I/O-pin being on the top or bottom side, and the other is an output I/O-pin being on the left or right side. The normal cases, an input I/O-pin on the left or right side and an output I/O-pin on the bottom or top side, can be handled as if the I/O port were a block.

The connections to the I/O-pins are illustrated in Figure 4.6. Only the top and left sides are shown; the bottom and right sides being symmetrical.

All the base I/O-pins are still on m_B, covering at least the height/width of a row/column. The input I/O-pin on the left side (PO:a) and the output I/O-pin on the top side (PI:U) work as usual.

The output I/O-pin (input I/O-port) on the left side (PI:V) connects to the vertical m_{B+2} trunk through a horizontal m_{B+1} wire, as if it is an input pin on the left side. This horizontal wire is called a pseudo-branch. The column of the virtual output pin or the trunk is chosen as the column into which the average X-coordinate of the input pins on this net falls, minimizing the associated branch lengths. The exact X-coordinate of the trunk will be determined by the result of the interval packing of that column. The following pseudo code summarizes the above operation:

Handling_Left_Output_I/O_Pin(po)

c=0 (assuming left I/Os in column 0), *nip*=0
for (each input pin *pi* of the net) {
   c+=Column(*pi*), *nip*++.
}
trunkColumn=*c*/(*nip*+1).
create pseudo branch in Row(*po*), from column 0 to column *trunkColumn*.
normal Fishbone construction.

The input I/O-pins (output I/O ports) on the top side (PO:b, PO:c or PO:d) use a so-called extended row to first create its virtual input pin, and then connect to its trunk through a horizontal m_{B+1} branch. The extended row does not have a *GR* in the Y direction. It extends vertically until all the virtual input pins have their horizontal branches accommodated via interval packing. The following pseudo code gives a summary of the handling of the top I/O input pins. The reason for giving the added vertical segments 0 X-grids is to prevent overlaps between the added segments and the trunks stretched into the extended row.

Handling_Top_Input_I/O_Pins

collect all branches for the top input I/O pins in the extended row.
  do a normal interval packing in the extended row for the branches with unlimited tracks.
for (each top input I/O pin *pi*) {
  stretch the trunk to Y(branch of *pi*).
  add a vertical segment [X(*pi*)^0, Y(branch of *pi*)~Y(*pi*),m_{B+1}] to connect the pin and the branch.
}

An extended column could be used for the output I/O-pins on the left side, but then the virtual output I/O-pin and its vertical trunk would appear in this extended region. Thus the input pin(s) on the corresponding net may all need to stretch their branches to the far left side of the layout. If there are multiple input pins, this is obviously bad for wire length/delay, unless all blocks carrying these input pins happen to be placed on the left side of the layout.

4.2.5. The Fishbone placement and routing flow

Our block placement uses a simulated annealing framework with sequence pair as the layout representation [MURA96]. Sequence pair is a representation of a non-overlapping block placement. A sequence is an ordered list of the blocks, imposing the spatial relationship (X or Y) between any pair of the blocks in the list. Given a pair of sequences, one for the right-up relation and the other for the left-down relation, the relative position of any two blocks is known. The blocks can be then packed in X and Y directions to produce the layout. Other block packing techniques can also be used [HONG00, YOUN03]. At each annealing step, a layout modification, such as block swapping (by swapping two blocks in one of the sequences) or orientation change, is made. Note that a block in the Fishbone scheme can only have four orientations: normal, X-flipping, Y-flipping and XY-flipping; 90 degree rotations are not allowed. Then the interval packing algorithm is run for each row and column to arrange branches and trunks, and the wire delays are computed. The cost function is

$$\begin{aligned} cost &= w_A \cdot A + w_{DA} \cdot D_A + w_{DM} \cdot D_M \\ &+ w_V \cdot N_V + w_{ASP} \cdot \max(0, ASP - ASP_0) \end{aligned},$$

in which A is the chip area, D_A is the average delay, D_M is the maximum delay, N_V is the number of violations (segments), and ASP is the chip aspect ratio. Parameters w_A, w_{DA} , w_{DM}, w_V and w_{ASP} are the corresponding weights, and

ASP_0 is a threshold above which the aspect ratio is penalized. In practice, the weights are scaled such that the variables with different units are normalized. The normalization is done as follows. Area A_0 and delays D_{A0} and D_{M0} are measured from the initial layout that is randomly generated. Then w_A is multiplied by $1/A_0$, w_{DA} by $1/D_{A0}$ and w_{DM} by $1/D_{M0}$. The goal of the algorithm is to seek a Fishbone placement with N_V=0 while the weighted sum of area, average delay, maximum delay and aspect ratio penalty is minimized. The choices of the weights depend on the design requirement. The cost function is flexible. The commonly used total (or average) wire length and other parameters can also be included in the cost function. We use delay instead of wire length in our experiments, because delay is usually the direct goal a designer pursues while wire length can be regarded as an approximation to wire delay. It is reemphasized that all the variables in the equation except N_V will receive exact values when a placement is given and the Fishbone nets are constructed. The flow is summarized in the following pseudo-code.

FishboneFlow

```
randomly generate an initial layout
for each scheduled annealing step {
    randomly do one of the following:
        (1) swap a pair in one of the sequence pairs
        (2) swap a pair of I/Os
        (3) flip one of the blocks
    update layout
    interval packing and counting violation number
    evaluate area, aspect ratio and delays
    evaluate cost
    accept or reject
}
```

As a comparison, we consider placements using MST net model. In the cost function, D_A and D_M are measured based on the construction of the nets with MST topology. But they are still estimated values, because no detailed routing is done. And MST for a net may not be unique, adding to the difficulty of making precise estimation.

We discuss experimental results without using buffer insertion in the next sub-section, while the more realistic delays are dealt with in Section 4.3 where buffer insertion is performed during placement and routing.

4.2.6. Experimental results

For testing full-chip block placement, three MCNC benchmark examples for block placement [MCNC], ami33, ami49 and playout, were used. In addition, 15 random block placement examples were generated. For the MCNC circuits, we moved each pin in a block to its nearest legal Fishbone location and made it into a base-pin (a stripe rather than a point). The random circuits were generated using as blocks 18 ISPD benchmark circuits for partitioning [ISPD98]. The 18 circuits, ibm01 to ibm18, formed a pool of blocks. Chip-level designs were generated by randomly selecting blocks (with replacement) from the pool; their interconnections were randomly generated. The interconnections followed a distribution that the average pin number per net was 3 while the maximum pin per net was 10.

Table 4.1: The characteristics of the testing examples

	# block	# IO-port	# net	# pin
ami33	33	42	117	522
ami49	49	22	407	953
playout	62	192	1609	4656
ibmB10	10	30	878	2634
ibmB15	15	45	1306	3918
ibmB20	20	60	1764	5292
ibmB25	25	75	2206	6618
ibmB30	30	90	2651	7954
ibmB35	35	105	3079	9239
ibmB40	40	120	3512	10536
ibmB45	45	135	3943	11830
ibmB50	50	150	4335	13006
ibmB55	55	165	4737	14211
ibmB60	60	180	5326	15980
ibmB65	65	195	5768	17304
ibmB70	70	210	6627	19881
ibmB75	75	225	7272	21818
ibmB80	80	240	7705	23115

The characteristics of the chip-level circuits are listed in Table 4.1. The technology information is listed in Table 4.2. The block size of each ISPD circuit was determined by $n_C \times (115\mu m^2)/0.8$, in which n_C is the cell count given by the benchmark, $115\mu m^2$ is the average cell size and 0.8 accounts for the area utilization (total cell area divided by block area) of the block. The pins of the blocks were randomly placed on the block boundaries at legal Fishbone positions. All the base pins are on layer m_B. Such pin placement should not affect the placement algorithm in comparison with the Fishbone design flow.

Table 4.2: Technology information

	unit	m_{B+1}	m_{B+2}
wire width	μm	0.8	0.8
wire spacing	μm	0.7	0.7
wire pitch	μm	1.5	1.5
direction		horizontal	vertical
area capacitance	pF/μm^2	4.3×10^{-6}	4.3×10^{-6}
fringe capacitance	pF/μm	6.1×10^{-5}	6.1×10^{-5}
square resistance	Ω/□	5.5×10^{-5}	5.5×10^{-5}
via resistance	Ω	1.0	
output pin res.	Ω	239.0	
input pin cap.	pF	0.035	

We assume that registers are attached to the pins of the blocks in the examples, so paths start from output pins (drivers) and terminate at input pins (loads). Two kinds of delays were measured in the experiment. One is the average delay, and the other is the maximum delay.

The Fishbone scheme was compared with an approach using MST during placement. The reason why an HP model was not compared with is that HP only gives wire length estimation and we want wire delays. The same simulated-annealing and sequence-pair placement was used [MURA96] for the MST; the only difference was just the different net models adopted. An iterative Steiner Tree heuristic was adopted in the placement [KAHN92] when using MST. The Steiner tree constructions were done at each random move of the annealing. Intuitively the MST approach should give the best placement result, although the run time can be very long. In the cost function, w_A=0.5, w_{DA}=0.3, w_{DM}=0.2, while w_V=0.03 was used for Fishbone and w_V=0.0 for MST, and w_{ASP}=0.05 with ASP_0=1.5. In the real design, the values of w_A, w_{DA} and w_{DM} can vary to meet the design requirement. The value of w_V for Fishbone was empirical. For all the testing circuits, the number of violations of Fishbone dropped to zero within the first few annealing temperatures with a w_V ranging from 0.1 to 1.0. In the MST based block placement, no routing congestion was estimated, so w_V was constantly zero. Since the base pins are stripes instead of points, the center points of the base pins were used to estimate the wire lengths and delays in the MST placement. This should not introduce significant error in estimating delays for MST; each pin can have at most $GR/2$ error in its position. The sizes of most circuits are over $100GR \times 100GR$; so the error is small for nets spanning many columns and rows. For nets spanning within a small region, the relative error of pin position is large. However in such cases, the wire capacitance and resistance are dominated by pin capacitance and resistance. Therefore using center points of the stripes to estimate delays for MST does not produce significant errors in delays.

When placement terminates, the Fishbone scheme has already produced the final layout. The block placement generated by MST was passed to the Cadence Warp Router (*Silicon Ensemble* Version 5.3) for the final routing. The Warp Router is also Steiner Tree based. The MST results after routing are labeled by MST*. Elmore model was used in the experiment to calculate delays [ELMO48]. The wires were segmented by pins, vias and "T" junctions. Each segment was viewed as a π type RC component. All programs were run on a Sun Blade 1000 workstation.

The results are given in Table 4.3. All Fishbone placement tasks terminated with no violations. On average, the Fishbone scheme resulted in a 5% area overhead, 5% increase in average delay and 10% increase in maximum delay, compared to the MST placement (without real routing). When the MST placement was finally routed, denoted by MST*, the area did not change, and the average and maximum delay differences became only 1% and 2%. However in two designs (ibmB45 and ibmB75), the routing of MST* did not complete successfully. The violations reported by Warp Router included two types. One is that some nets were unrouted. The other is that all nets were routed but overlaps existed. In the latter case, each overlap counted as one violation. If a wire of a failed net crosses many other wires, the violation number could be large. This is exactly what happened in reporting violations of the two failed examples. For the first case, the failed nets were simply excluded from delay computation; while in the second case, overlaps were ignored and the wire delays were computed as usual. In reality, the design flow must be repeated, if any violation occurs. The area and delay overheads of the Fishbone scheme can be considered as a (small) price paid for routability and predictability. Warp Router reported, in the problematic cases, no over-capacity G-Cells (a coarse routing grid used in the global routing to estimate possible wiring congestion). Most of the routing violations occurred in small pin regions, meaning that such routability problems are hard to predict even in the early routing stages. Leaving more space between blocks and/or utilizing more routing layers may help, but it is unclear if, or how much, such extra resources should be allocated. We observed that sometimes the router built two or more connection points to a base pin; such flexibility is not used or required in the Fishbone scheme. However, this did not seem to help in solving the routability problem.

Table 4.3: Experimental results

	area (mm^2)		delay_average (ps)			delay_max (ps)			#violation		run time (minute)		
	FB	MST	FB	MST	MST *	FB	MST	MST *	FB	MST *	FB	MST	MST-tot
ami33	1.71	1.74	148	161	169	653	715	743	0	0	0.2	5.2	5.3
ami49	43.2	44.7	267	251	271	836	837	927	0	0	0.8	13.1	13.2
playout	118	111	316	346	365	713	750	856	0	0	7.1	46.1	46.8
ibmB10	47.5	51.5	501	477	500	927	887	968	0	0	4.8	45.2	45.7
ibmB15	72.8	72.3	521	534	558	1060	959.1	1004	0	0	7.6	54.3	55.2
ibmB20	99.1	95.2	601	574	602	1306	1176	1283	0	0	12.0	76.1	77.6
ibmB25	135	129	703	640	675	1490	1297	1333	0	0	15.5	95.6	98.0
ibmB30	150	146	681	681	717	1430	1354	1498	0	0	19.4	117	120
ibmB35	182	175	746	721	751	1830	1361	1442	0	0	25.6	85.4	90.1
ibmB40	211	198	760	735	764	1669	1507	1624	0	0	35.0	109	114
ibmB45	240	230	819	786	816	1887	1746	1824	0	132	44.4	138	146
ibmB50	270	249	835	792	824	1913	1682	1740	0	0	48.8	167	173
ibmB55	291	289	870	845	877	2108	1843	1996	0	0	70.7	209	216
ibmB60	365	303	950	868	902	2125	1945	2065	0	0	85.7	253	260
ibmB65	370	339	1053	880	916	2358	1822	2038	0	0	96.3	283	292
ibmB70	481	395	1039	915	952	2242	1988	2332	0	0	73.5	318	327
ibmB75	547	536	1144	1019	1064	3077	2218	2826	0	9k	86.0	359	418
ibmB80	666	541	1260	1066	1107	3060	2347	2706	0	0	91.2	661	675
compare	1.0	.95	1.0	.95	.99	1.0	.90	.98			1.0	6.6	6.77

The run times of the Fishbone and MST schemes are the times taken only by the simulated annealing phase. Fishbone is on average 560% less than for MST placement; the MST needs extra time for routing. If the total run time (placement and final routing) of the MST approach is considered, the Fishbone scheme is 577% faster. Other Steiner Tree algorithms might reduce the run time for MST placement [BORA94], but the average pins per net of all the testing examples are smaller than 5. The speed-up by using a faster Steiner Tree algorithm might be limited. The MST designs with routing failures may need even more time to complete the flow.

Wire length results (average wire length, which is total wire length divided by net number) are given in Table 4.4, although wire delays were the main concern in the experiment. On average, Fishbone gives 11% longer wire length. For circuit ibmB65, Fishbone produced 34% longer wire length than MST*, while in terms of average and maximum delays, the overheads were only 15% and 16%. The main reason is that only the average and maximum delay terms appeared in the cost function, while wire length was not an optimization goal. On the other hand, there might be some correlation between wire length and delays. Circuit ibmB65 is an example where longer wire length usually means larger delay. Circuit ami49 provides a counterexample. Fishbone produced longer wire length and smaller delays. The experiment has also shown that smaller wire length may not even give better routability (ibmB45 and ibmB75).

Table 4.4: Wiring comparison

	wirelength (mm)		#segment		#via	
	FB	MST*	FB	MST*	FB	MST*
ami33	0.48	0.58	293	909	524	803
ami49	3.03	2.48	1055	4316	1866	3656
playout	4.93	5.26	4285	18643	7500	16303
ibmB10	7.76	6.32	3255	14682	5280	12876
ibmB15	7.71	7.82	5240	22942	7854	20169
ibmB20	9.47	8.37	7086	31572	10620	27957
ibmB25	12.6	10.2	8858	41432	13286	36877
ibmB30	11.8	11.1	10648	47894	15958	42483
ibmB35	12.9	12.2	12366	55624	18534	49592
ibmB40	13.7	12.2	14103	56119	21140	49074
ibmB45	15.0	13.1	15851	64568	23756	55915
ibmB50	15.2	13.4	17408	71450	26104	61921
ibmB55	16.0	14.5	19032	75567	28532	64985
ibmB60	18.2	14.9	21388	87689	32072	75757
ibmB65	20.6	15.4	23157	93160	34720	80204
ibmB70	20.5	16.2	26600	113250	39890	97275
ibmB75	22.6	18.3	29198	146762	43790	127920
ibmB80	26.0	19.4	30923	131129	46380	110012
compare	100%	89%	100%	424%	100%	240%

An important result is the number of wire segments and vias of the final routing, because these two parameters shed some light on the "geometric simplicity" of the routing layers. It is quite reasonable to speculate that designs with less wire segments and vias are more favorable for the manufacturing process. At least such designs are easier to apply mask engineering techniques like OPC. The results show that the Fishbone scheme, compared to MST*, has 324% less segments and 140% less vias. The number of segments and vias also affect the run time of the delay computation. A simple net topology like the Fishbone even makes possible a hard coded program for delay computation, which can reduce run time significantly. Unpredictable net topologies like the MST require a general delay computing engine. More segments and vias slow down the computation.

Table 4.5: Difference of the delays of MST and MST*

	delay_average (ps)			delay_max (ps)		
	MST	MST*	Δ	MST	MST*	Δ
ami33	161	169	5.0%	715	743	9.1%
ami49	251	271	4.4%	837	927	4.7%
playout	346	365	5.0%	750	856	9.1%
ibmB10	477	500	5.0%	887.2	967.7	9.1%
ibmB15	534	558	4.4%	959.1	1004	4.7%
ibmB20	574	602	5.0%	1176	1283	9.1%
ibmB25	640	675	5.5%	1297	1333	2.8%
ibmB30	681	717	5.3%	1354	1498	10.6%
ibmB35	721	751	4.2%	1361	1442	6.0%
ibmB40	735	764	4.0%	1507	1624	7.8%
ibmB45	786	816	3.9%	1746	1824	4.4%
ibmB50	792	824	4.1%	1682	1740	3.4%
ibmB55	845	877	3.8%	1843	1996	8.3%
ibmB60	868	902	3.9%	1945	2065	6.2%
ibmB65	880	916	4.1%	1822	2038	11.8%
ibmB70	915	952	4.1%	1988	2332	17.3%
ibmB75	1019	1064	4.4%	2218	2826	27.4%
ibmB80	1066	1107	3.9%	2347	2706	15.3%
average			4.7%			9.6%

Finally we investigate the accuracy of the MST as an estimation model in the placement by comparing the estimated delay values and the actual values (routed). Table 4.5 gives the differences between the estimated average and maximum delays during the placement (MST) and the real average and maximum delays after routing (MST*). It is no wonder that both delay values are underestimated during the placement, 4.7% and 9.6% for average and maximum delays respectively, because the routability issues or routing congestion are not taken into account. Recall that the global routing stage of Warp Router did not report congestion in the designs. Therefore estimating routing congestion during the MST placement may most probably give no useful information either. It is possible to introduce certain margins, 9.6% for instance, in the delay estimation during the placement. But this may affect the area, since delays and area are additive components in the cost function. Even if a margin as large as 9.6% were set, there would still be some examples showing errors bigger than 9.6%, and the errors themselves are not predictable.

Table 4.6: Difference of the individual net delays of MST and MST*

	ave.	std.	max	max	percentage		
	Δ	dev. Δ	Δ+	Δ−	Δ+	|Δ|<1	Δ−
ami33	5%	6%	14%	>−1%	100%	0%	0%
ami49	8%	9%	34%	>−1%	99%	1%	0%
playout	6%	7%	29%	>−1%	100%	0%	0%
ibmB10	5%	6%	24%	>−1%	99%	1%	0%
ibmB15	4%	5%	24%	>−1%	98%	2%	0%
ibmB20	5%	6%	38%	>−1%	99%	1%	0%
ibmB25	5%	7%	42%	>−1%	100%	0%	0%
ibmB30	5%	7%	33%	>−1%	99%	1%	0%
ibmB35	4%	6%	29%	-6%	94%	4%	2%
ibmB40	4%	5%	29%	-7%	94%	4%	2%
ibmB45	4%	5%	31%	-8%	93%	4%	3%
ibmB50	4%	5%	28%	-7%	94%	4%	2%
ibmB55	4%	5%	28%	-8%	93%	5%	2%
ibmB60	4%	5%	36%	-6%	93%	5%	2%
ibmB65	4%	6%	39%	-8%	94%	4%	2%
ibmB70	4%	6%	36%	-9%	94%	4%	2%
ibmB75	5%	8%	121%	-8%	93%	4%	2%
ibmB80	4%	5%	37%	-7%	94%	4%	2%
average	5%	6%					

The estimation errors may not be uniform for all nets. This effect is examined in Table 4.6, in which the statistics of the delay estimation errors of individual nets are given. On average, a net can cause a 5% underestimation (ave. Δ), and the standard deviation (std. dev. Δ) of the errors is 6%. The maximum underestimation (max. Δ+) and overestimation (max. Δ–) are also given. It is interesting to see that actually some delays were overestimated. This is possible, because when a net is detoured to prevent congestion, some pin-pin connections are lengthened while others shortened. In addition, the center points of the stripe shaped base pins were used to build the MST during the placement, while the router may make different but slightly better choices for the connection points on the base pins. Although the maximum under/overestimation errors can be as large as 121%, the corresponding nets may not happen to be the ones with the maximum delays. The last three columns show the percentages of the wire delays that were underestimated (Δ+), correctly estimated (|Δ|<1), and overestimated (Δ–). The majority of the delays were underestimated.

Figure 4.7 shows the layout pictures of "playout", which is designed with the Fishbone scheme.

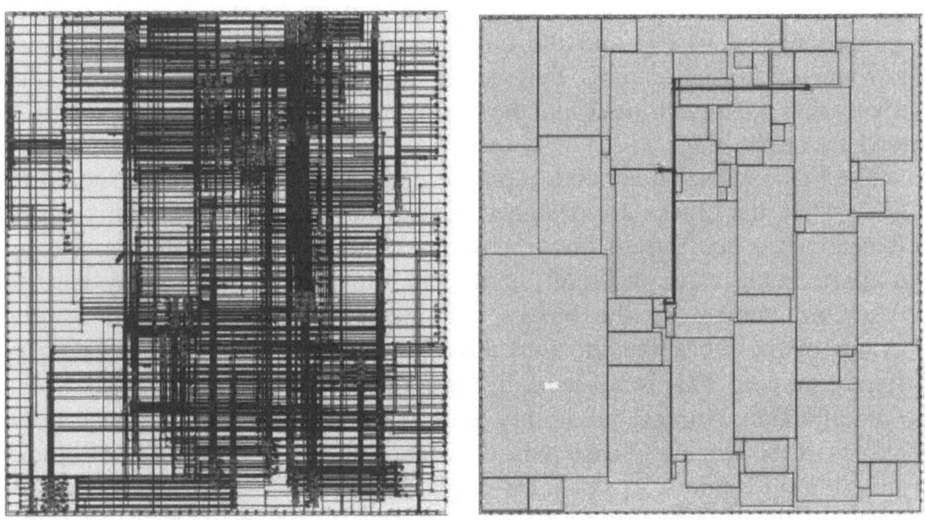

(a) the layout

(b) the block placement and one four-pin net highlighted

Figure 4.7: The layout of "playout"

4.2.7. Summary of the Fishbone scheme

In this section the Fishbone block-level placement and routing scheme was presented. The fixed and simple net topology enables fast wire construction and precise wire delay evaluation, which makes possible easy integration of placement and routing. The configurability of pin location using base/virtual pin pairs offers great flexibility in placing wires and making connections. Thus routability is very often achievable even with only two routing layers. The cost is some loss of optimality in area and delay. For instance, block orientations are reduced from eight kinds to four (no 90° rotations), which might cause the elimination of some better placements. The wire delay of Fishbone is usually (very slightly) inferior to that of the MST. But if the placement made with MST is eventually un-routable, as the experimental results show, then their superiority in area and wire delay is diminished.

The current version of the Fishbone cannot handle obstructions in the Fishbone routing layers. In block integration, obstructions on global routing layers are atypical. However if some blocks are sensitive to signals running above like the analog modules, designer may wish obstruction regions built into the global routing layers above these blocks. In general, this is not compatible with the Fishbone scheme. If the chip can be sliced into two parts, one for sensitive blocks and the other for regular blocks, we can still use Fishbone for the latter part. This seems to be feasible, because it is very unlikely that sensitive blocks like the analog modules are allowed to be placed anywhere on the chip.

The Fishbone scheme needs a pre-defined radix *GR*. The radix should be known when the blocks are designed. Using the radix formula in this paper, this becomes a question of how many blocks are to be integrated on a chip. An approximate value is usually sufficient. However, if it is really hard to predict the number of the blocks, the block designer may offer several versions of the block layout, with different base pin positions according to different radixes. This is not hard, because the core layout of the block needs no modification. Another possibility is to introduce local routing between the real pins of the block and base pins. Consider the case of integration of blocks from different vendors, in which the original pins of blocks are not "Fishbone-aware" at all. After determining a common *GR*, we can route each pin to the nearest valid base pin location on layer m_B. This might not be hard, especially when the original pin spacing is larger than *GR* routing grids. We have used a similar technique when experimenting with the MCNC benchmark examples.

Although the Fishbone scheme does well in reducing the wire delays during the block placement and routing, the delays may still be too large to meet timing requirements. The delay of an un-buffered wire is roughly a quadratic function of the wire length, but good buffering may bring the wire delay back to a linear function of the wire length. So it is not only worthwhile but also necessary to consider buffer insertion in the block level placement and routing. A Fishbone scheme with integrated buffer insertion function is presented in the next section.

4.3. Fishbone Scheme with Buffer Insertion

Block-level placement and routing is quite different from that at the gate-level, in the following three aspects [YOSH01, CONG01a]:

(1) The block-level placement is a hard problem in both theory and practice [MO00, SAIT99, SHER93]. Fortunately the numbers of blocks in most designs are within the range of several tens or hundreds. Simulated-annealing based algorithms usually can generate layout with little white space at the expense of run time [SHER93].

(2) The wiring density in block-level routing is not as high as in gate-level. However, block-level routing usually makes use of higher metal layers, and the wires must connect to the pins of the blocks which are on lower layers, producing routability problems, in and around pin regions in particular. Such routability problems are hard to detect unless detailed routing is performed.

(3) Block-level routing deals with longer interconnections, which forces consideration of wire delays. As feature sizes get smaller, the wire effect becomes a dominant part in the total delay [MO03a]. Buffer insertion is a commonly adopted approach for fixing timing problems [CONG01a, CONG01b, HU02], as described in Section 2.2.4.2. However, buffer insertion after routing may be limited by the areas in which buffers are allowed to be placed. The connections to the buffer pins also introduce routability problem, because wires on the higher layers must connect to the bottom layers to reach the buffers and then connect back up.

Each aspect alone may find good or acceptable solutions. However, in reality we have to face all of them, and they are inter-related. Due to the typical long run-times of simulated-annealing, precise but slow estimation of routability is not affordable during the placement. For the same reason, buffer insertion is made a separate procedure, or some rough estimation is done early. These lead to design non-convergence, or "timing closure" problems.

In this section, the basic Fishbone integrated placement and routing scheme is enhanced by buffer insertion (Fishbone-B or FB-B). Besides the buffer insertion, a method requiring large blocks to embed buffers is proposed. Since some blocks in the layout are relatively large, like the cache modules in microprocessors, wires running across them may find no place to insert buffers unless some buffers have already been embedded in these blocks. Of course the white spaces between blocks are also feasible locations for buffers, and these do not require modification of blocks. All the buffers, whether in blocks or white space, are laid out with respect to the Fishbone grid rules. As a result, their pins are like the block pins, which always locate on certain specific routing grids such that they are easily reachable by wires on higher layers. Given a placement, the pin locations as well as the wires are known for all nets because of the spine topology; also the locations of all buffers are known; therefore the buffer insertion problem reduces to choosing buffers from known positions to minimize the wire delays.

This section is organized as follows. In Section 4.3.1, two types of buffers are discussed. Section 4.3.2 gives a buffer insertion algorithm. Section 4.3.3 outlines the Fishbone-B flow. Section 4.3.4 gives experimental results, and 4.3.5 concludes the section.

4.3.1. The buffers

We only consider uniformly sized non-inverted buffers. There are two kinds used in the Fishbone scheme; one is the buffers embedded in the blocks (E-Buffers), the other is the buffers placed in the white space of the layout (W-Buffers). E-buffers are pre-built in the blocks, hence they are fixed. W-buffers simply fill up the white space in the layout. A buffer occupies the area of $GR \times GR$ grids. With practical design rules, a buffer can be constructed easily in such area if $GR \geq 5$. But even if $GR > 5$, we still use buffers with size of 5×5 grids. Of course in such cases, buffer sizing, especially for W-buffers, becomes another option. With respect to orientation, buffers can be further classified as "branch" buffers and "trunk" buffers, as shown in Figure 4.8.

A "branch" buffer is vertically laid out, with input on the bottom and output on the top. The base pins for a "branch" buffer are located on the left and right. If a branch comes from left to right (the input pin of the branch is on the right and the trunk of the net is on the left), the left base pin connects to the input pin on the bottom while the right base pin connects to the output pin on the top. If the branch comes from right to left, the pin pairings are reversed.

In this way a "branch" buffer can handle either a branch towards the left or right. A "trunk" buffer is similar, except its base pins are on the bottom and top. The "branch" and "trunk" W-buffers are interleaved in a checkerboard fashion. The E-buffers are organized into rows of "trunk" buffers and columns of "branch" buffers. At the cross points of the rows and columns, there needs to be one more E-buffer location to bridge the gaps.

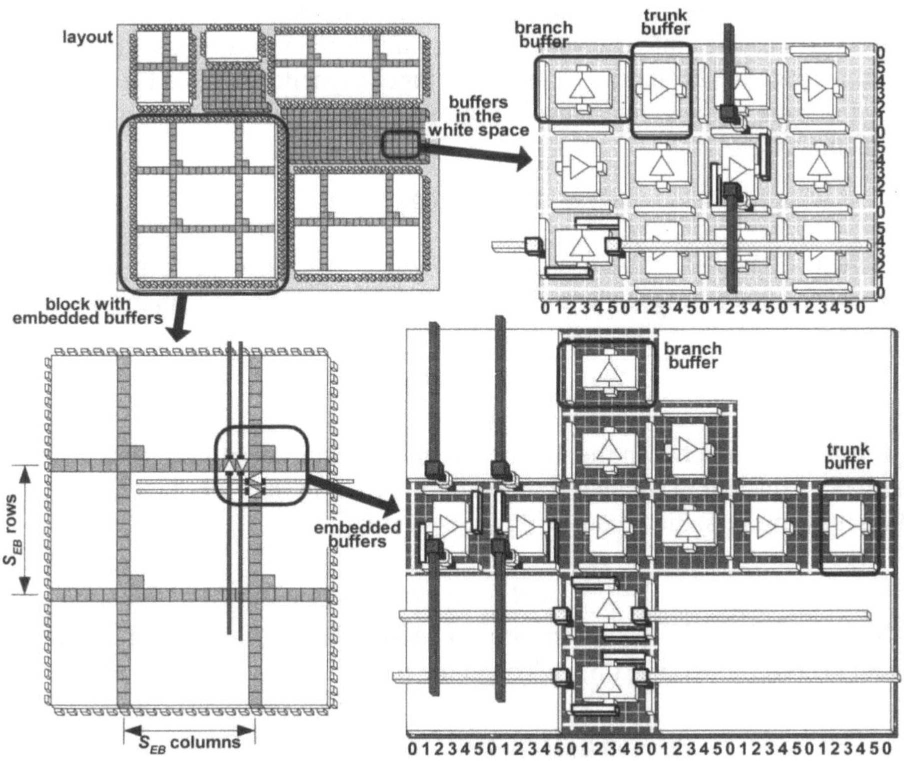

Figure 4.8: The W-buffers and E-buffers.

An important parameter in the above buffer organization is the E-buffer row/column spacing, denoted by S_{EB} (in units of *GR*-rows/columns). In fact each block in a design can have different S_{EB}'s, and also horizontal and vertical directions can have different S_{EB}'s, but we only consider the case that the same S_{EB} is used for all. Note that small blocks may contain no E-buffers. Obviously, smaller S_{EB}, i.e. more E-buffers, provides more opportunities for buffer insertion, and different nets will have less competition for buffer resources inside the block regions. However, denser E-buffers mean larger block sizes, which in turn increases wire lengths and makes more wires require buffering. The derivation of an optimal S_{EB} is hard and S_{EB} can be different from circuit to circuit. Various versions of a block could be designed that contain E-buffers organized with different S_{EB}'s. However, this task is not as easy as creating versions for different pin GR's, because the quantity of E-buffers affects the area of the block and thus the wire lengths of the internal connections. A reasonable estimation of S_{EB} can be made as follows. Consider a pin-to-pin connection (fanout=1) of a wire with length of L. Suppose that N_B buffers are inserted into the wire, dividing the wire into (N_B+1) segments with equal length. The buffers have the same input capacitance C_I as the input pins and the same driving resistance R_O as the output pin. The change of delay (reduction) is:

$$\Delta d = N_B\left[\frac{L^2 R_{//} C_{//}}{N_B+1} - (R_O C_I + D_{BUF})\right],$$

in which, $C_{//}$ and $R_{//}$ are the unit length wire capacitance and resistance, and D_{BUF} is the load independent delay of the buffers. To maximize the delay reduction,

$$N_B(L) = L\sqrt{\frac{R_{//}C_{//}}{R_O C_I + D_{BUF}}} - 1 \cdot$$

A linear relation between the number of inserted buffers and the wire length is derived. Consider a branch horizontally crossing the whole chip layout. Without E-buffers, the layout width is L_0. In total

$$N_1 = \frac{L_0}{GR \cdot Pitch(m_{B+1}) \cdot S_{EB}}$$

columns of "branch" E-buffers are embedded in the blocks and the layout width becomes L_1=L_0(1+1/S_{EB}). To get the best buffering effect, $N_1 \geq N_B(L_1)$, or

$$\frac{L_0}{GR \cdot Pitch(m_{B+1}) \cdot S_{EB}} \geq L_0\sqrt{\frac{R_{//}C_{//}}{R_O C_I + D_{BUF}}} - 1 \cdot$$

The "−1"on the right side is dropped for simplicity. The above estimation was made for one wire. Assume half of the wires in a row or column are long wires that need buffering,

$$S_{EB} = \frac{2}{GR^2 \cdot Pitch(m_{B+1})}\sqrt{\frac{R_O C_I + D_{BUF}}{R_{//}C_{//}}} \cdot$$

With the technology parameters in the experiment,

$$S_{EB} = \frac{1.45\times 10^4}{GR^2} \cdot$$

This indicates that when GR=10 (the width of a column is 15μm), a column of E-buffers are embedded per 145 columns, if the width of a block contains at least 145 columns (or 2.1mm). This also shows that E-buffers only occupy a very small portion of the block area (about $1/S_{EB}^2$), which eliminates the worry that embedding buffers in blocks may affect the intra-block timing.

4.3.2. Buffer insertion

Given a placement of the blocks, a bit map *BufAva*(*r*,*c*) can be quickly built, indicating whether at row=*r* and column=*c* a buffer is available. A bit is tri-valued, standing for "no or unavailable buffer", "W-buffer" or "E-buffer".

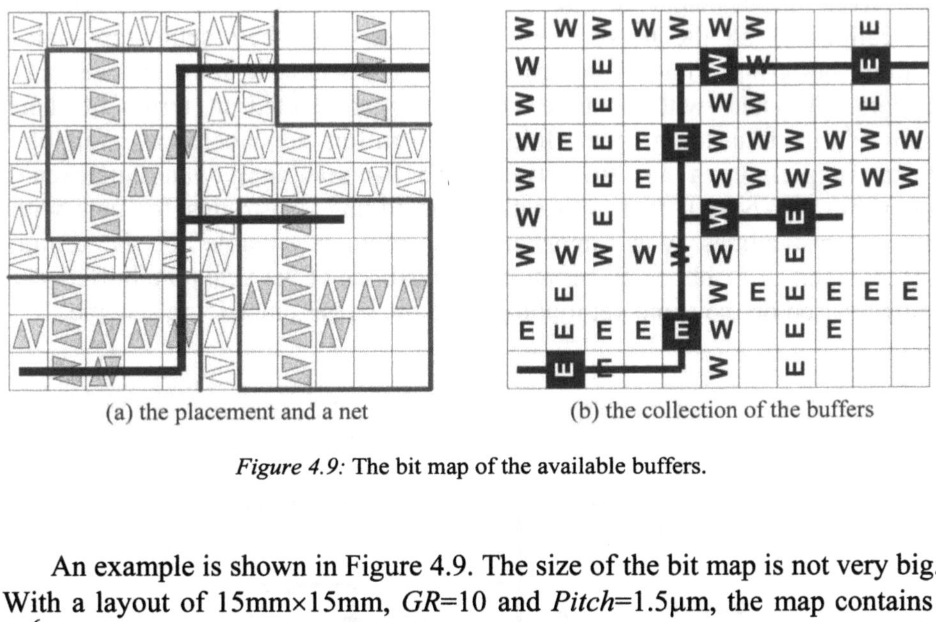

(a) the placement and a net (b) the collection of the buffers

Figure 4.9: The bit map of the available buffers.

An example is shown in Figure 4.9. The size of the bit map is not very big. With a layout of 15mm×15mm, *GR*=10 and *Pitch*=1.5μm, the map contains 10^6 bits. The spine net topology makes a net easily identify which (*row*,*column*)'s its wires cover, and thus the number of buffers available to the net and their locations are known. Note that buffers have orientations, so "branch" buffers are only available to branch wires and "trunk" buffers are only available to trunk wires. A net can choose a subset from the buffers available to it to minimize the net delay.

Different nets may compete for the buffer resources. Therefore an arbitration mechanism is needed to determine the sequence of the nets for buffer consideration. Define function *Cr*(*net*) as the criticality of a net. The function outputs zero if the net is non-critical or a positive value representing the criticality. A unique feature of this algorithm is that no net criticality needs to be defined before placement. Unlike most other algorithms in which *Cr*(*net*)'s are pre-defined numbers, the criticalities are computed based on the real net topologies and wire lengths during the Fishbone placement and routing. This eliminates the possibility of wrong assignment of the net criticalities before placement due to lack of useful physical information. The formulation of *Cr*(*net*) is based on the slack:

$$Cr(net) = -\min\{0, \min_{ipin \in net}(t_{REQ}(ipin) - t_{ARR}(ipin))\},$$

in which, the difference of the required time t_{REQ} and the arrival time t_{ARR} of the input pin is the slack. The definitions of these can be found in Section 2.3. The criticality of a net is always a non-negative number. If *Cr*(*net*)=0, then the constraint on this net is satisfied, so the net does not need buffer insertion.

Given a placement of the blocks and the routing of the nets, the criticalities of all the nets are measured based on their spine topologies without buffering. Thus if the criticality of a net is 0, it does not need any buffers. The nets with positive criticalities are sorted in terms of their *Cr*(*net*)'s in descending order. Initially all buffers are labeled as available. Next we pick up a net from the sorted list and apply van Ginneken's algorithm [VANG90] to try to satisfy the timing constraints for nets with positive criticalities. Van Ginneken's algorithm is suitable for the timing optimization of Fishbone-B, because the locations of the available buffers are known based on the spine net topology and the buffer embedding scheme.

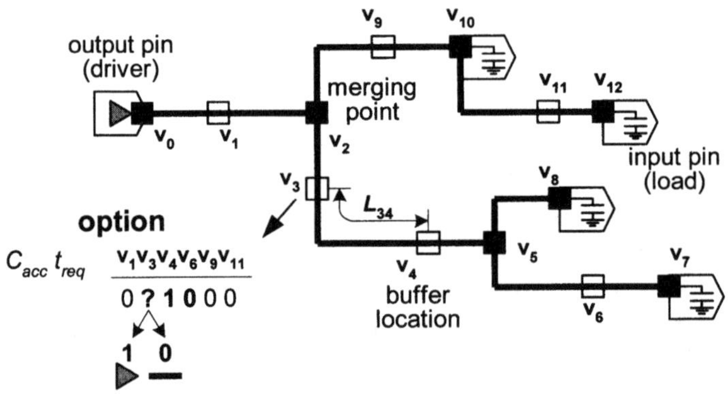

Figure 4.10: Applying van Ginneken's algorithm on a buffering tree.

We give a brief description of van Ginneken's algorithm, using the example in Figure 4.10. In the delay analysis, the Elmore model is adopted [ELMO48]. In fact the discussion here is more general than the original form of the algorithm reported in [VANG90]. A net can be viewed as a tree with the root being the output pin (driver) and all leaves being the input pins (loads). Note that an input pin can be an intermediate node, as v_{10} in the figure. The black squares in the figure are either pins or the so-called "merging points", which simply join two or more edges. The white squares are the buffer locations, which are always on the edges. At each buffer location, we can choose to insert or not insert a buffer. The term "option" is a triple, containing the accumulated capacitance C_{acc}, the required time t_{REQ} and a bit vector V. For input pins, $t_{REQ}(ipin)$'s are exactly the required times set in the timing constraint (see Chapter 5 for how this is determined in a particular application). The bit vector records whether buffer insertion takes place at each buffer location. A "0" means no buffer (i.e. continuing the wire) and a "1" means an inserted buffer. The algorithm selects the best option that satisfies the timing constraint. The satisfaction of the timing constraint is tested by $t_{ARR}(opin) \leq t_{REQ}(opin)$, where $t_{ARR}(opin)$ is the arrival time of the output pin given in the timing constraint.

The algorithm is recursive, starting from the root (output pin) and traversing the entire tree. The operation on a node depends on the type of the node:

(1) At a leaf node (i.e. an input pin), one option is generated: $\{C_{ipin}(ipin), t_{REQ}(ipin), \text{all-0}\}$.

(2) At a merging point, the sets of options of all children are combined. Consider the case of two children, labeled $child_1$ and $child_2$. Each of the children has a set of options, describing the candidate options of the sub-tree rooted at the child node. Denote the sets by $OS(child_1)$ and $OS(child_2)$. The merging combines each pair of options from the two sets, $o_1 \in OS(child_1)$ and $o_2 \in OS(child_2)$, using:

$$C_{acc}(new) = C_{acc}(o_1)+C_{acc}(o_2)+C_w(child_1)+C_w(child_2)+C_?$$
$$t_{REQ}(new) = \min[t_{REQ}(o_1)-d_w(child_1),\ t_{REQ}(o_2) - d_w(child_2)]$$
$$V(new) = V(o_1) \text{ or } V(o_2),$$

in which, C_w and d_w are the wire capacitance and delay between the child and the current node, and $C_?$ is the input pin capacitance if the merging point is also an input pin (as v_{10} in the example) or zero. The wire delay d_w has the form of $d_w=R_w \times (C_w/2+C_{acc})$, where R_w is the wire resistance. Both R_w and C_w are linear functions of the wire length. Although the number of the new options generated through merging can be as many as $|OS(child_1)| \times |OS(child_2)|$, a pruning process may reduce the number significantly. A new option is obviously worse, if both its C_{acc} is greater and its t_{REQ} is smaller than those of another new option. Such new options are removed immediately.

(3) At a buffer location, there are two choices, continuing the wire or inserting a buffer. The former requires that each option o of the option set of the child be inherited but with the accumulated capacitance and required times updated:

$$C_{acc}(new) = C_{acc}(o)+C_w(child),$$
$$t_{REQ}(new) = t_{REQ}(o)-d_w(child),$$
$$\text{and } V(new) = V(o).$$

If a buffer is inserted,

$$C_{acc}(new) = C_{ipin}(buffer)$$

and the best option o_{BEST} of its child is inherited:

$$t_{REQ}(new)=t_{REQ}(o_{BEST})-d_w(child)-R_{opin}(buffer)\times[C_{acc}(o_{BEST})+C_w(child)],$$

in which, $R_{opin}(buffer)$ and $C_{ipin}(buffer)$ are the output resistance and input capacitance of the buffer. An option $o \in OS(child)$ is the best option, if

$$t_{REQ}(o)-R_{opin}(buffer)\times C_{acc}(o)$$

is the maximum. The new bit vector is:

$$V(new) = (00001000) \text{ or } V(o_{BEST}),$$

and the "1" is at the bit corresponding to this buffer location.

The recursive algorithm ends with a set of options, denoted by $OS(root)$, at the output pin. The options in the set are further processed to include the delay caused by the output resistance of the driver. Then the options that satisfy $t_{ARR}(opin) \leq t_{REQ}(option)$ will meet the timing constraint of this net. In contrast to van Ginneken's original algorithm, in which the option that gives the largest $t_{REQ}(opin)$ is chosen, we select an option that satisfies the timing constraint and makes use of the least buffers (counting "1" in V). When some buffers are selected by the algorithm, they become unavailable for the subsequent nets. The process is iterated for all the nets in the list.

The available buffers are collected for the net by tracing the trunk and branches in the $BufAva(r,c)$ bit map. Since the complexity of van Ginneken's algorithm is $O(\#B^2+\#IP)$ in which $\#B$ is the number of buffer locations and $\#IP$ is the number of input pins of the net, we need to be careful in controlling $\#B$. The E-buffers are no problem, because a wire usually meets the next E-buffer after S_{EB} columns or rows. However, it is very likely that W-buffers are clustered. Directly collecting the buffers along a wire may cause many closely located W-buffers to be included, which slows down the algorithm significantly. Actually, nearby buffers are very unlikely to be selected together. Therefore, we prevent collecting all the W-buffers in a cluster along a wire by setting a skipping-range denoted by S_{WB}. When a W-buffer is collected, it requires S_{WB} columns or rows to collect another W-buffer in the same W-buffer cluster (between these W-buffers are all W-buffers). The following pseudo code shows the operation, but only the collection of available buffers for the lower half of the trunk is detailed.

```
collect_buffer(net n)
B=Φ.
for (r=row(opin),c=col(opin),wBuf=0; r>min(row(ipin's)); r-=1) {
   if (AvaBuf(r,c)="unavailable") continue.
   else if (AvaBuf(r,c)="E-trunk-buffer") B=B∪{buffer(r,c)}.
   else if (AvaBuf(r,c)="W-trunk-buffer") {
      if (wBuf=0) B=B∪{(r,c)}
      if (wBuf++ =S_WB) wBuf=0.
   }
}
for (r=row(opin),c=col(opin),wBuf=0; r<max(row(ipin's)); r+=1) {
   collect available trunk-buffers on the upper half of the trunk.
}
For (each input pin ipin) {
   collect available branch-buffers on the branch.
}
return B.
```

The buffer insertion procedure is summarized in the following pseudo code:

```
buffer_insertion
compute Cr(n) based on the spine nets without buffers.
sort all the nets with Cr(n)>0 in list L.
#timing_violation=|L|.
for (L≠Φ) {
   n=next net from L.
   B=collect_buffer(n).
   SetOfOption=vanGinnekenAlgorithm(n,B).
   o*∈SetOfOption, that o* uses minimum number of buffers
                        and makes Cr(n)=0.
   if find an o* {
      for (buffer∈o*)
         BufAva(row(buffer),column(buffer))="unavailable".
      #timing_violation-=1.
   }
}
return #timing_violation.
```

4.3.3. The Fishbone-B physical design flow

Our block placement uses a simulated annealing framework with sequence pairs as the layout representation [MURA96]. At each annealing step, a layout modification, such as block swapping or change in orientation, is made. Note that a block can only have four orientations: normal, X-flipping, Y-flipping and XY-flipping; 90 degree rotations are *not* allowed. Then the interval packing algorithm is run for each row and column to arrange branches and trunks, and the wire length is computed. The cost function is:

$$cost = w_A \cdot A - w_D \cdot \min_{ipin}\lfloor 0, t_{REQ}(ipin) - t_{ARR}(ipin) \rfloor + w_R \cdot \#V_R,$$

where A is the layout area, the second term account for the most negative slack, and $\#V_R$ is the number of routing violations. Parameters w_A, w_D and w_R are the corresponding weights. The goal of the algorithm is to seek a Fishbone placement with $\#V_R=0$ while the weighted sum of the first two terms minimized. The design flow is summarized in the following pseudo code.

```
randomly generate an initial layout.
for each scheduled annealing step {
    randomly do one of the following:
        (1) swap a pair in one of the sequence pairs, update layout.
        (2) swap a pair of I/Os.
        (3) flip one of the blocks.
    interval packing and counting violation number.
    buffer insertion.
    evaluate area, delay violations and routing violations.
    accept or reject
}
```

Interval packing is very fast, because it usually handles only a limited number of branches/trunks in a row/column. Therefore it can be run in the inner loop of the simulated annealing. The routing congestion is fully defined by the Fishbone routability condition and is reflected by the number of routing violations. Buffer insertion is based on the current placement and routing; its quality is reflected by the number of delay violations. The design flow terminates when the placement is finished; no separate routing and buffer insertion are necessary, because the wires have been finalized using the Fishbone topology with the buffers inserted.

4.3.4. Experimental results

The fifteen artificial examples in Section 4.2 were tested, which were created by randomly choosing blocks from the eighteen ISPD benchmark examples and building their interconnections. The arrival times set on the output pins and the required times set on the input pins were also randomly generated. The technology information is given in Table 4.7, and m_B corresponds to metal 2. The routing layers are metal3 (m_{B+1}) and 4(m_{B+2}). The examples were tested with the following configurations: (1) MST-none: Steiner Tree net topology and no buffers, (2) MST-W: Steiner Tree net topology with only W-buffers, (3) FB-B-none: Fishbone with no buffers, (4) FB-B-W: Fishbone with W-buffers only and (5) FB-B-W&E: Fishbone with both W-buffers and E-buffers. Block sizes became larger as buffers were embedded. For Steiner Trees, after the placement and buffer insertion, we fixed the buffer locations and called the Cadence Warp Router to complete the routing. This might cause routing violations. The Fishbone-with-Buffer-Insertion (FB-B) completed the placement, buffer insertion and routing in one single run.

Table 4.7: Technology information

Layer	metal2	metal3	metal4
width (μm)	0.6	0.8	0.8
pitch (μm)	1.2	1.5	1.5
C_{AREA} (pF/μm^2)	4.0×10^{-5}	2.0×10^{-5}	2.0×10^{-5}
C_{FRINGE} (pF/μm)	1.0×10^{-4}	1.0×10^{-4}	1.0×10^{-4}
R_{SQUARE} (Ω)	8.0×10^{-4}	8.0×10^{-4}	8.0×10^{-4}
$R_{VIA-down}$(Ω)	1.0	1.0	1.0
preferred direction	none	vertical	horizontal
C_I(pF)	0.0059		
R_O(Ω)	100.0		
D_{BUF}(ps)	11.0		

The net criticality formulation in Section 4.3.2 was used. In all configurations, the weighted sum of the chip area and the Most Negative Slack (MNS) is minimized. In the simulated-annealing, w_A=0.5, w_D=0.5 and w_R=0.03. Of course it is rare in practice that Steiner Trees are built and buffer inserted at each annealing step. However the costly runs give a good reference for the results of the FB-B algorithm. All programs were run on a Sun Blade 1000 workstation.

Table 4.8: Area and delay

	area (mm^2)					-MNS (ps)				
	MST		FB-B			MST		FB-B		
circuit	none	W	none	W	W&E	none	W	none	W	W&E
ibmB10	50	50	50	48	48	544	65	577	66	61
ibmB15	72	71	73	70	72	662	70	678	76	65
ibmB20	99	98	103	100	103	743	78	755	79	73
ibmB25	129	128	130	133	133	880	79	882	81	77
ibmB30	148	149	152	151	158	985	87	1004	88	79
ibmB35	180	178	185	184	188	1102	87	1117	93	80
ibmB40	217	209	211	210	211	1239	90	1266	93	83
ibmB45	240	234	240	233	239	1377	89	1401	93	85
ibmB50	258	254	259	252	260	1509	95	1510	95	89
ibmB55	309	310	312	313	327	1643	112	1640	111	100
ibmB60	357	360	355	354	372	1788	122	1782	123	108
ibmB65	408	406	414	413	428	1904	122	1931	123	109
ibmB70	449	453	453	456	477	2073	129	2100	131	114
ibmB75	550	549	560	562	565	2146	140	2195	140	122
ibmB80	671	683	677	679	694	2239	147	2418	148	128
compare	1	0.99	1.01	1.00	1.03	1	0.079	1.02	0.080	0.072

The area and delay results are given in Table 4.8. Routing violations and run times are reported in Table 4.9. Most of the configurations show very similar layout area, except that the FB-B-W&E configuration has on average 3% area overhead. Both MST-W and FB-B-W reduce MNS by a factor of about 92%, compared with their non-buffer configurations. This shows the great power of buffer insertion in reducing delay. Note that the buffers in these two configurations make use of existing white space. MNS can be further reduced by an additional 1% (from 0.080 to 0.072) if E-buffers are available in the FB-B case. But the reduction is at the cost of a 3% area increase due to the embedded buffers. Fishbone-B achieves 100% routability in all examples, while in a few examples MST fails in the routing. When the routing violations occur (such as wires crossing each other), we still compute the delays as if the wires are valid.

Table 4.9: Routing violations and run time

	#routing violation					run time (hour)				
	MST		FB-B			MST		FB-B		
circuit	none	W	none	W	W&E	none	W	none	W	W&E
ibmB10	0	0	0	0	0	0.7	1.6	0.1	0.4	0.4
ibmB15	0	0	0	0	0	0.7	2.5	0.1	0.7	0.7
ibmB20	0	0	0	0	0	0.9	2.9	0.2	0.9	0.9
ibmB25	0	0	0	0	0	1.3	3.9	0.3	1.0	1.0
ibmB30	0	0	0	0	0	1.5	4.8	0.3	1.3	1.3
ibmB35	2	0	0	0	0	1.7	5.4	0.4	1.3	1.4
ibmB40	0	0	0	0	0	2.0	5.7	0.5	2.0	2.0
ibmB45	0	0	0	0	0	2.3	7	0.7	2.2	2.4
ibmB50	0	0	0	0	0	2.9	7.2	0.9	2.3	2.5
ibmB55	0	0	0	0	0	3.8	7.7	1.0	2.5	2.6
ibmB60	0	0	0	0	0	5.0	8.6	1.1	2.5	2.9
ibmB65	0	0	0	0	0	6.2	9.2	1.2	2.6	3.3
ibmB70	0	0	0	0	0	7.9	9.6	1.2	2.8	3.5
ibmB75	10	33	0	0	0	8.6	9.7	1.4	3.4	3.8
ibmB80	0	0	0	0	0	9.9	10	1.5	3.6	4.1
compare						1	2.4	0.20	0.69	0.73

It is interesting to find that even with embedded buffers, which cause block sizes to grow, the total layout areas of some examples are the same as those of the FB-B-W configuration (ibmB10 for instance), and even smaller than the FB-B-none configuration (ibmB10 for instance). One reason is that the E-buffers only occupy a small portion of the block area. The other reason is that these buffers are very helpful in reducing wire delays, making the annealer concentrate more on area minimization.

In Table 4.10, the buffer utilizations of the different configurations are shown. In general only a small portion of the W-buffers are used to speed up the circuits. However when E-buffers exist, they have more chances of being utilized by the buffer insertion algorithm. The W-buffers, although free to use, are not always present where the buffer insertion algorithm wants most. The E-buffers, owing to their regular distribution, make it easier for the buffer insertion algorithm easier to find buffer locations which are more helpful in reducing wire delays. The E-buffers are not free though. Therefore controlling the number of E-buffers through S_{EB} is an important design tradeoff that affects the chip area and the wire delay reduction. Further experiments on the choice of S_{EB} and other E-buffer schemes are needed to refine this choice.

The layout of ibmB10 produced by FB-B-W is shown in Figure 4.11.

Table 4.10: The utilization of the buffers
number of available buffers (used%)

	MST	FB-B		
	W	W	W&E	
circuit	W-buf	W-buf	W-buf	E-buf
ibmB10	105205 (1%)	760 (1%)	94941 (1%)	945 (14%)
ibmB15	111147 (1%)	1222 (1%)	105342 (1%)	1308 (14%)
ibmB20	126383 (1%)	1823 (2%)	152486 (1%)	2447 (23%)
ibmB25	86125 (3%)	2264 (2%)	139402 (1%)	2795 (14%)
ibmB30	155689 (2%)	2806 (2%)	170707 (1%)	3812 (9%)
ibmB35	170152 (2%)	3317 (2%)	220156 (1%)	4180 (6%)
ibmB40	189631 (2%)	3852 (3%)	200649 (2%)	5284 (8%)
ibmB45	180385 (2%)	4419 (2%)	183747 (2%)	4203 (9%)
ibmB50	146368 (3%)	4981 (4%)	139894 (2%)	5039 (9%)
ibmB55	285993 (2%)	6438 (2%)	299808 (2%)	7343 (8%)
ibmB60	418592 (2%)	8020 (2%)	422216 (2%)	8446 (6%)
ibmB65	419874 (2%)	8437 (2%)	483441 (2%)	9702 (5%)
ibmB70	424864 (3%)	12560 (3%)	445954 (2%)	11133 (6%)
ibmB75	505741(4%)	17571 (4%)	525893 (2%)	16599 (7%)
ibmB80	523752 (4%)	19579 (4%)	503911 (4%)	18578 (8%)
compare	(2.3%)	(2.4%)	(1.7%)	(10%)

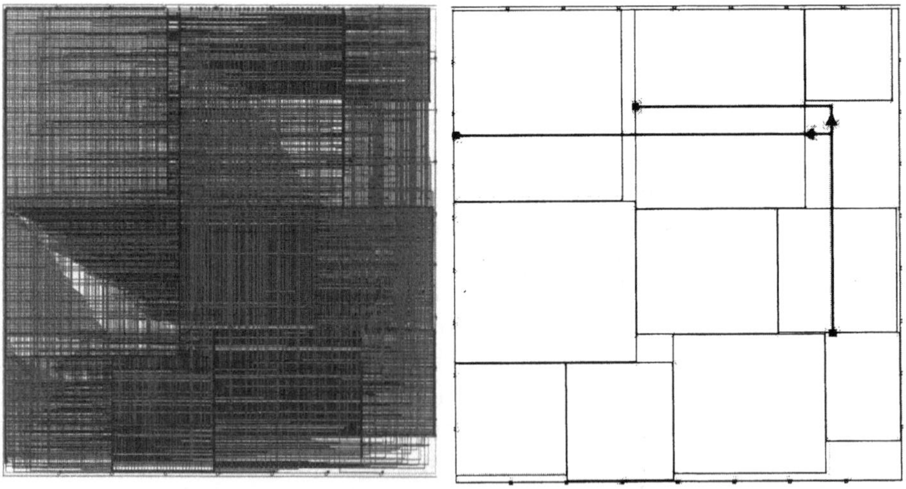

(a) the layout

(b) the block placement and one buffered net with buffers shown

Figure 4.11: The layout of ibmB10 (W-buffers only).

4.3.5. Summary of the Fishbone-B

The Fishbone block-level placement and routing scheme with buffer insertion was presented. The fixed and simple net topology enables fast wire construction and precise wire delay evaluation, which makes possible integration of placement, routing and buffer insertion in a single simulated-annealing framework. Net criticality, an important factor in timing-driven physical design, can be computed in "real-time". A buffer embedding scheme was also proposed, which is consistent with the column/row configuration of the Fishbone method. All buffers, W-buffers and E-buffers, have a Fishbone style pin configuration, which alleviates the routability problem for the connections to the buffers. Experimental results show that the use of embedded buffers, along with buffers placed in the white space, greatly improves wire delays.

Following are some problems worth discussing. (1) The presence of the W-buffers is not controlled. If the layout is too perfect with little white space, the number of W-buffers may become very limited, weakening the power of buffer insertion. It might be interesting to consider techniques like buffer block planning [CONG01b] to provide better control of the W-buffers. (2) The buffers may be sized also, within a $GR \times GR$ area, to get even smaller wire delays. (3) In block-level designs, especially microprocessors, many nets are grouped into buses. With the Fishbone scheme, the nets in a bus are routed with very similar shapes. Buffer insertions for them are also very similar. So a bus can be treated as a single net to save run time. An advantage of the Fishbone scheme is that the adjacent nets of a bus are separated by at least one routing track (the 0-grid), incurring lower coupling noise than those going closely side by side.

4.4. Summary

In this chapter, an integrated block-level placement and routing scheme was presented. The Fishbone scheme makes use of a simple and regular spine topology, which offers good predictability. The idea of base-virtual pin pairs and the introduction of cyclic grid allow easy and fast construction of the routing of the whole layout. This also enables the integration of the placement and routing in a single simulated-annealing framework. Wire delay measured directly from the Fishbone routing can appear in the cost function. Moreover, the regular net topology reduces the numbers of wire segments and vias, indirectly enhancing manufacturability. In the second half of the chapter, buffer insertion was added to the basic Fishbone scheme. The Fishbone-B scheme can simultaneously analyze wire delay and plans buffers, owing to the predictable spine net topology. An embedded buffer scheme was discussed also, enhancing the availability of the buffers and further improving timing. The Fishbone algorithm provides comparable results in terms of area and delay to Steiner Tree based algorithms, but it runs much fast. The Fishbone-B scheme is used at the physical design stage in the Module-Based design flow, which is to be presented in the next chapter.

Chapter 5

THE MODULE-BASED DESIGN FLOW

5.1. Overview

In this chapter, a Module-Based full-chip design flow is presented. The chapter starts with a detailed discussion of the timing issues and the delay/area relationship in Section 5.2, which forms the basis of the timing-driven design flow. In Section 5.3, the Physical Synthesis flow is discussed. This flow has been widely accepted by the IC industry, because it takes physical information into account at the very early stages of the flow. However, as addressed in Section 2.5, DSM challenges make this seemingly sound approach far from guaranteeing timing closure within a reasonable design time. Our new design flow is given in Section 5.4, which is based on the concept of different versions of modules. The versions of modules are a kind of flexibility generated at the early stages of the flow and utilized by the later stages. The main idea is to eliminate as much as possible unreliable estimations. The Module-Based design flow is compared with the Physical Synthesis flow, and the results are reported in Section 5.5. Section 5.6 summarizes.

5.2. Basics of the Design Flow

In this section, some basic factors of the design flow are discussed. In 5.2.1, the concepts of module and path are reviewed, because these are the fundamental elements in timing driven design. The input to a design flow is a netlist containing hard modules and soft modules. Also a target clock cycle is provided. Constraints for soft modules are generated based on the external interconnection of the modules. We use the conventions of "required times" and "arrival times" to establish the timing constraints for the modules, described in 5.2.2. The constraints affect the internal structures of the modules. The direct results are the module areas. In 5.2.3, the relationship between area and constraints is analyzed, which also re-emphasizes how the unpredictable area/constraint tradeoff may jeopardize timing closure. 5.2.4 summarizes the problems and possible solutions.

5.2.1. Modules and paths

Synchronous digital circuits with a single clock source are considered. A design can be specified as a set of modules that are inter-connected. Hard modules (HM), such as memory blocks or other IP blocks, are both logically and physically fixed. Soft modules (SM) contain only initial Register-Transfer-Level (RTL) descriptions and need logic synthesis and physical design.

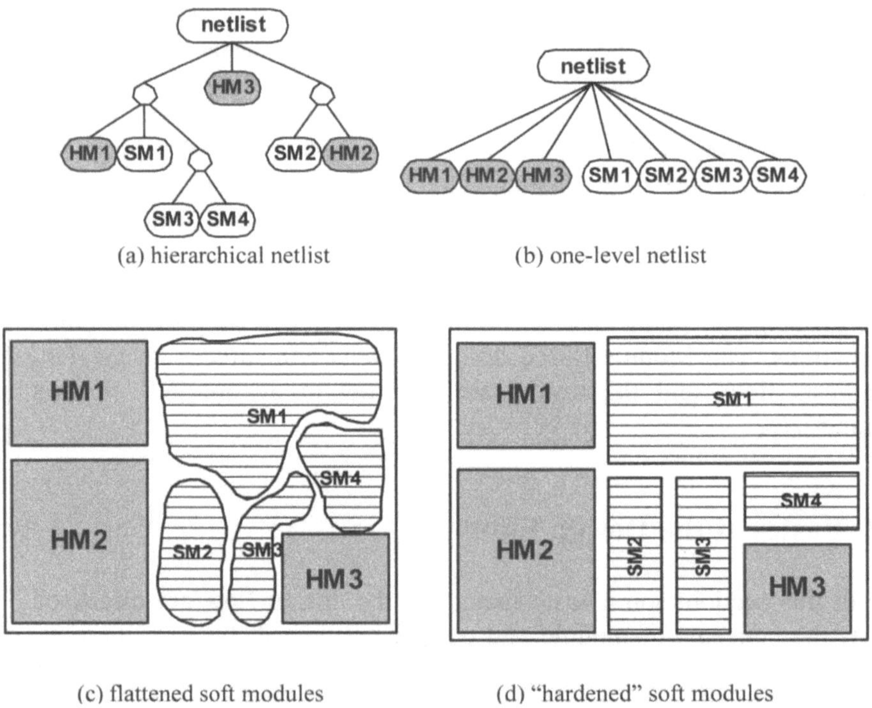

Figure 5.1: Hard modules and soft modules.

An example is shown in Figure 5.1. The netlist, as illustrated in Figure 5.1(a), contains 3 hard modules (HM1..3) and 4 soft modules (SM1..4); the modules are organized hierarchically. The hierarchy is removed and the modules are now on one-level, as illustrated in Figure 5.1(b). We work on the one-level netlist. One way of physically implementing the netlist is to synthesize each soft module and then flatten them in the layout, as shown in Figure 5.1(c). There is no clear boundary between the cells logically belonging to different soft modules, although the cells of the same soft module may be placed close to each other. The other approach is to maintain the soft modules after the synthesis, such that each soft module physically becomes a hard module, as illustrated in Figure 5.1(d).

The goal is to implement the input netlist with a small clock cycle time and small chip area with the clock cycle given higher priority.

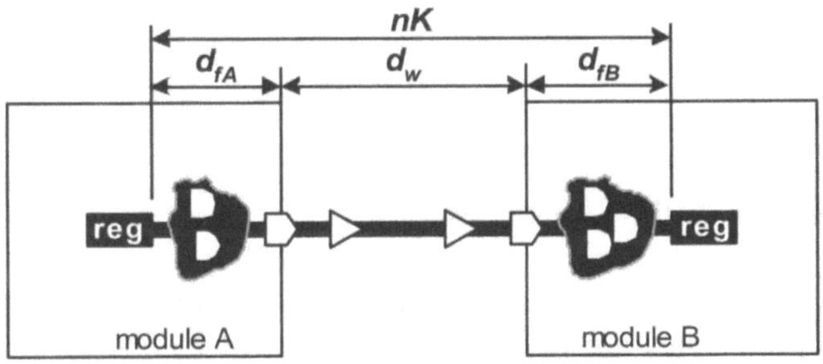

Figure 5.2: The path delay.

A combinational path, abbreviated "path", is a connection from one register to another (or the same) register, through logic gates and wires. As illustrated in Figure 5.2, a path delay consists of three parts, the internal delay in the driver module denoted by d_{fA}, the wire delay denoted by d_w and the internal delay in the load module denoted by d_{fB}. Setup and hold times of the registers are ignored for simplicity. Wires can be buffered to speed up signal propagation. If the driver and load modules are the same, we call the path an "internal path". If all the soft modules are flattened such that the design only contains one large soft module, the number of internal paths can be huge.

Denote path delay by $d(path)$ and the number of clock cycles allowed for the signal propagation by $n(path)$. For a given clock cycle K,

$$\frac{d(path)}{n(path)} \leq K$$

must be satisfied for all paths. For asynchronous connections, $n(path)=\infty$. Obviously,

$$\max_{path}\left\lceil \frac{d(path)}{n(path)} \right\rceil$$

determines the minimum working clock cycle of the chip, denoted by K_W. In the following discussion, we will only use single-cycle paths for simplicity. Usually the desired clock cycle K_0 is given as a starting point. However, when the flow cannot meet this, two options are present. One is to increase the clock cycle and run the flow again. The other is to modify the input netlist (possibly the behavior of the circuit) and run the flow again with the same clock cycle. The latter is beyond the scope of our discussion. We focus on the former; hence the problem can be stated as follows: Starting from clock cycle K_0, find the smallest working cycle $K_W \geq K_0$ such that all path delays are satisfied, and the chip area is reduced. Run-time is used as a criterion for the efficiency of the design flow; if a flow cannot find a solution for a given clock cycle, it should terminate early so that a new run can be started with an adjusted clock cycle.

5.2.2. Timing constraints for the modules

It is convenient to represent the timing constraints by "required/arrival times". Required times are defined on the pins of the modules and register inputs. The real times when the signals propagate to those nodes, known as "arrival times", must be no later than the required times. The difference between the required time and arrival time of a pin is called the "slack". Negative slack means a timing violation. Constraints in the form of arrival and required times are the input to a synthesis tool.

Hard and soft modules differ in the setting of the timing constraints, as shown in Figure 5.3. Inside a module, hard or soft, is a sequential circuit composed of registers and combinational logic. The internal logic of a hard module is fixed; so the output pins carry arrival times accounting for known internal delays; similarly the input pins carry required times. For soft modules, the register-to-output-pin and input-pin-to-register delays are the result of the synthesis. A soft module is said to be fast if such delays are short; slow otherwise. This can be controlled by setting arrival times on input pins and required times on output pins. There is also an implicit set of constraints which set arrival times to zero on all register outputs and required times K on all register inputs.

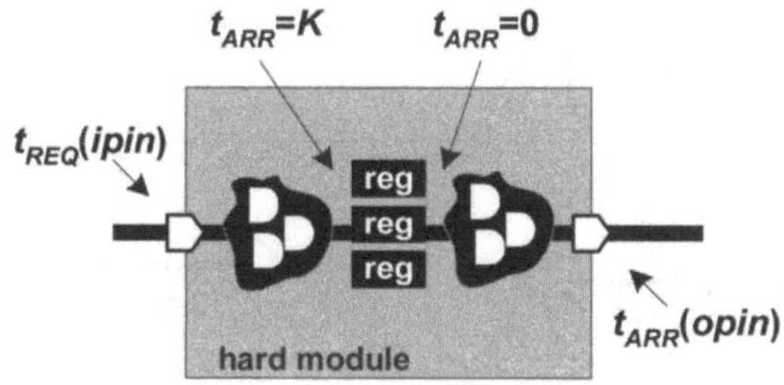

(a) hard module

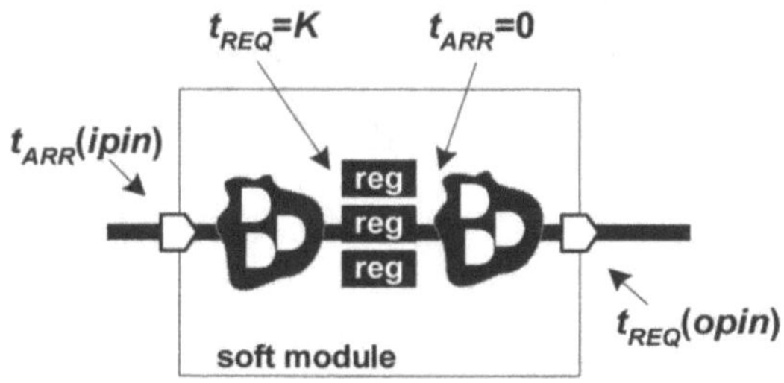

(b) soft module

Figure 5.3: The timing constraints set on the pins of the module.

A hard module already has a minimum working clock cycle, $K_W(hm)$:

$$K_W(hm) = \max\{d(path_{internal}), d_f(reg, opin), d_f(ipin, reg)\}.$$

It is the maximum delay among all its internal paths, register-to-output-pin combinational paths and input-pin-to-register combinational paths. Of course the maximum $K_W(hm)$ among all hard modules gives a lower bound for K_W:

$$K_W \geq \max_{hm} K_W(hm).$$

This can serve as a good starting clock cycle (K_0) for the flow.

The phrase "arrival/required times on pins of a module" is vague. Figure 5.4 shows an example.

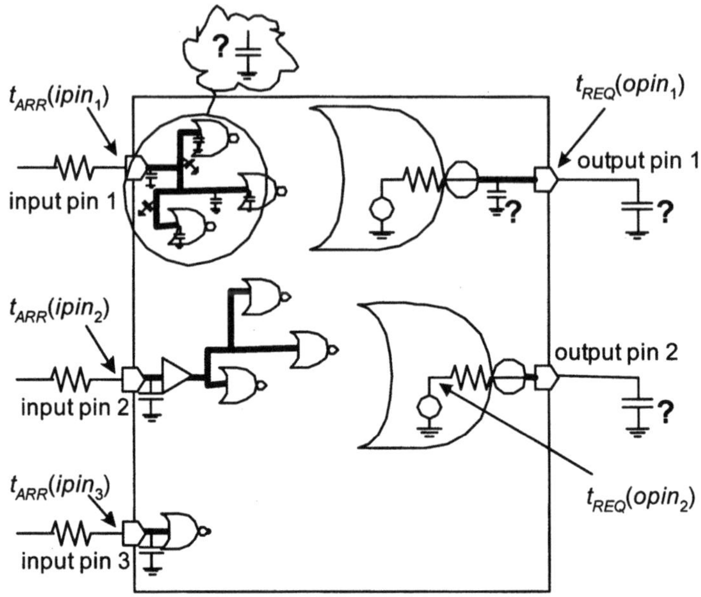

Figure 5.4: The arrival and required times, revisited.

Input pin 1 is internally connected to multiple gates. The number of gates connected to the pin is not known before the synthesis of the module. The wires between the pin and the gates contribute unknown capacitances, and the exact arrival time at the pin is a function of these capacitances. This causes inaccuracy in interpreting the meaning of the arrival time in the synthesis. To avoid this vagueness, we assume that only one gate is directly connected to the input pin and it is placed close to the pin such that the connecting wire can be ignored. In the case of multiple gates connected to the input pin, a buffer may be inserted right at the position of the input pin and it fans out to the gates, input pin 2 in the figure for example. If only one gate is connected to the input pin, we just assume that the gate is placed close to the pin, input pin 3 in the figure for example. Technology mapping can handle the added requirement by setting input driving capacity to one load (the capacitance of one gate input pin). The placer may need to keep the buffer or single gate close to the pin. The situation at the output pin is different. There is only one driver gate. The vagueness here is that the total load to the output resistor of the gate is unknown during the synthesis. The total loading capacitance includes internal wire capacitance and external capacitance, as for output pin 1. The solution, output pin 2 for example, is to place the gate close to the pin and set the required time of the soft module at the point just before the driving resistance. This resistor drives external loads which are mostly wire capacitances; so it is convenient to treat its effect as part of the wire delay.

In the rest of the discussion, we will still use the old notations and labeling of arrival/required times without re-emphasizing the new convention just defined.

5.2.3. The relation of module area and constraints

The widely used area/delay curve is confusing, because the term "delay" here does not have a strict definition [TABB98]. Instead, we call it an area/constraint curve.

In Figure 5.5, an area/constraint curve is illustrated. The module in Figure 5.5(a) is abstract, having only one input pin and one output pin. The arrival time is set on the input pin and the required time on the output pin. The gates in the black region relate to the paths through the input and/or output pin. The gates in the grey region are independent of such paths. Hence different required times only affect the area of the black region, while the grey region and the registers contribute a constant area. The relation between the area and $t_{REQ}(opin)$, the required time on the output pin, is analyzed in Figure 5.5(b), with $t_{ARR}(ipin)$ set to a constant. If $t_{REQ}(opin)$ is too small such that logic synthesis cannot find a netlist to satisfy it, then the curve falls into the "no solution" region. The minimum feasible required time $t_{REQ\text{-}MIN}(opin)$ corresponds to the maximum area $area_{MAX}(opin)$. As the constraint is loosened, the area drops. But when $t_{REQ}(opin)$ exceeds $t_{REQ\text{-}MAX}(opin)$, the area saturates, corresponding to $area_{MIN}$. In Figure 5.5(c), the area/input-pin-constraint relation is shown. The input pin is given an arrival time, $t_{ARR}(ipin)$. However, for the input pin, the smaller the arrival time is, the looser the timing constraint. So the X-axis in this curve uses negative arrival time. The saturated areas of the two curves are the same, while the maximum areas may be different. A very important point is that both $t_{REQ\text{-}MIN}$ ($t_{ARR\text{-}MIN}$) and $t_{REQ\text{-}MAX}$ ($t_{ARR\text{-}MAX}$) are not known, unless exhaustive search is conducted. However $area_{MIN}$ might be quickly found using a synthesis with no constraints.

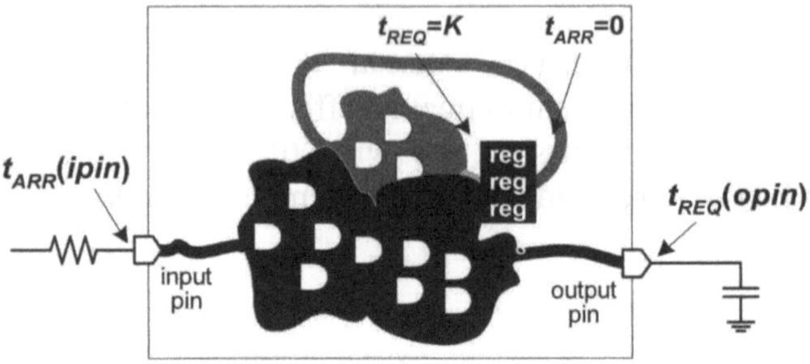

(a) the schematic

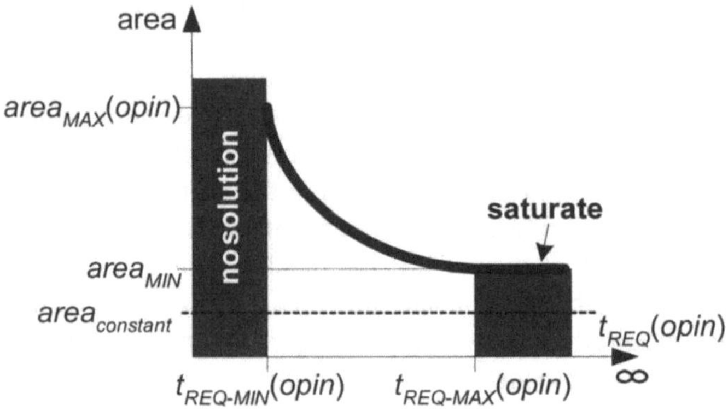

(b) area/output-pin-constraint curve

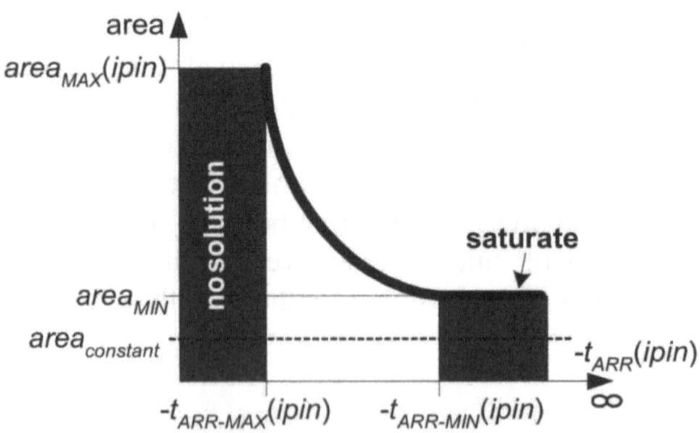

(c) area/input-pin-constraint curve

Figure 5.5: The area/constraint curve.

When the synthesis is pushed into the "no-solution" region by very tight constraints, it still produces results. Although the constraints are certainly unsatisfied, the netlist can be viewed as being "as-fast-as-possible" resulting in knowledge about $t_{REQ\text{-}MIN}$ and $-t_{ARR\text{-}MAX}$. The Module-Based design flow makes use of this feature.

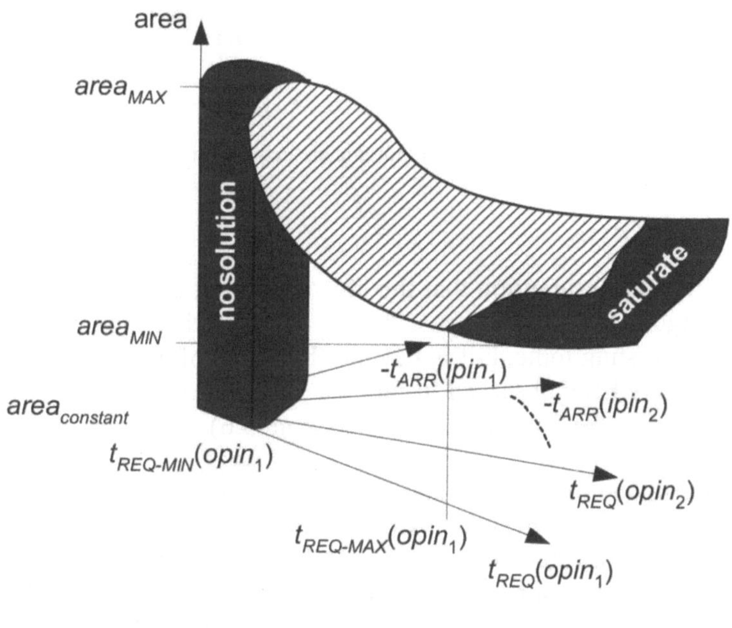

Figure 5.6: The area/constraints surface for a module.

In general, a soft module has many input and output pins. Each input pin is given an arrival time and each output pin a required time when the module is synthesized. Therefore the curve becomes a surface of area/constraints, as illustrated in Figure 5.6. In the following context, we may still call the multi-dimensional relation an area/constraint-curve for simplicity and draw it as a curve.

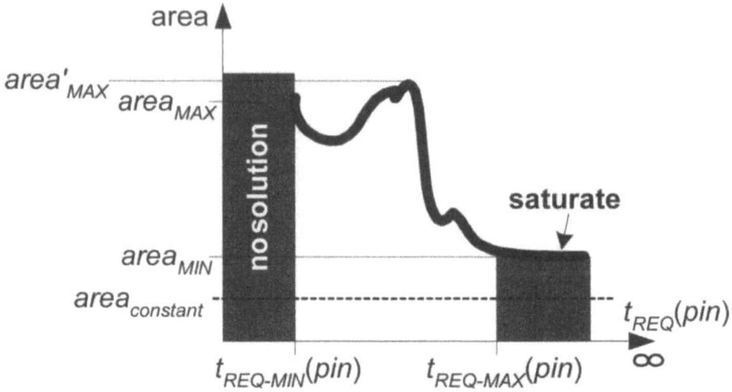

Figure 5.7: Non-monotonic area/constraint curve.

In reality, the assumption that the area/constraint curve is monotonic may not hold. Figure 5.7 shows a possible curve. Studying monotonic synthesis can be an interesting topic, but is beyond the scope of our discussion. Similar phenomena have been observed in the experimental results of River PLA and Whirlpool PLA (see Figure 3.29 in Section 3.6), although the area/delay curves are shown there.

The non-monotonic area/constraint relation is also a source of timing problems, especially in design flows with iterations between physical design and synthesis. Together with unpredictable wire delay, as addressed in Section 2.5.1.1, the unpredictable area/constraint relation makes timing closure a very hard problem. Synthesis of the modules is done with assumptions on wire delays. The predicted wire delays are transformed into timing constraints that are set on the pins of the modules to be synthesized. The synthesis tool produces a netlist for each module, satisfying the constraints, or reaching a point on its area/constraint curve. The discrepancies of the predicted wire delays and the real wire delays derived after physical design may require some modules to be re-synthesized with updated wire delay information. The operation of re-synthesis can be regarded as adjusting the locations of the points on the area/constraint curves. Convergence may be achieved if the physical structure before re-synthesis, i.e., the placement and routing of the modules, remains unchanged. Unfortunately maintaining such physical invariance is almost impossible, because re-synthesis of a module with a new set of timing constraints is very likely to change the area of the module, forcing changes in the placement.

5.2.4. The problems and possible solutions

From the above discussion, two key problems with modern IC design flow are captured:

(1) Wire delays are part of the path delays. These cannot be measured with much accuracy until routing is done. Unfortunately routing is the last step in the flow; in contrast, synthesis needs the wire delays so that timing constraints can be properly generated. In DSM designs, wire effects become more dominant; thus ignoring or badly estimating them during synthesis leads to problems. Synthesis works with timing constraints derived from wiring estimation. The estimation may be inaccurate. Physical design works with the synthesized (or further flattened) modules, expecting that the final wiring results are consistent with those estimated before synthesis. Unfortunately consistency is hard to guarantee. The synthesis or re-synthesis of a module may result in unpredictable module area, which can change the existing layout and undermine the fidelity of the wiring information derived from that layout. As the example in Figure 2.17 shows, the area change of one module affects not only its own wiring but also the wiring of other modules, further complicating the problem.

(2) The distribution of a path delay among the three parts, as shown in Figure 5.2, is another hard problem. Even if the wire delay can be safely ignored, how to split the path delay between the driver and load modules (both are soft modules) remains a question. This affects the setting of the arrival and required times on the pins and the modules and thus the synthesis results for these modules.

The two design flows to be described in the next two sections provide possible solutions: The Physical Synthesis approach, discussed in Section 5.3, tries to increase the accuracy of the pre-synthesis wiring estimation. The Module-Based design flow, presented in Section 5.4, takes a totally different approach. It tries to reduce as much as possible the need for pre-synthesis wiring prediction. Instead it creates flexibilities, which can be used in the physical design to adapt to various wiring situations.

5.3. The Physical Synthesis Flow

To solve the problem of inaccurate wire delay estimation before synthesis, the so-called Physical Synthesis design flow has been developed. It has become the most widely adopted timing-driven design flow [SYNO, CADE]. The main idea of Physical Synthesis is to floorplan the locations of the modules and their interconnections before (logic) synthesis starts.

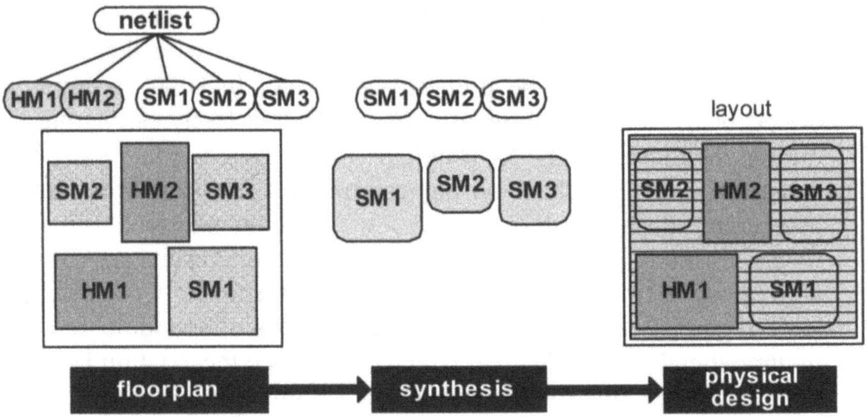

Figure 5.8: An example of the Physical Synthesis.

A simple example of the Physical Synthesis flow is shown in Figure 5.8. One run of the flow contains three stages: floorplanning, synthesis and physical design. The hard and soft modules are floor-planned such that their relative locations on the layout are determined. These locations are used to estimate wire delay and generate timing constraints. The synthesis tool produces the cell netlist for each soft module using derived constraints. When each soft module has its cell netlist determined, the physical design starts. Previously soft modules might be flattened such that their boundaries disappear, but ideally the cells belonging to the same previously soft module should be placed close to each other. The floorplan generated earlier is maintained, but some adjustment might be needed, because the area of the synthesized soft modules may be different from that used in the floorplanning. It is hoped that such adjustment causes no timing violations.

A complete flow should include check points for timing satisfaction and routability. A typical Physical Synthesis flow is shown in Figure 5.9.

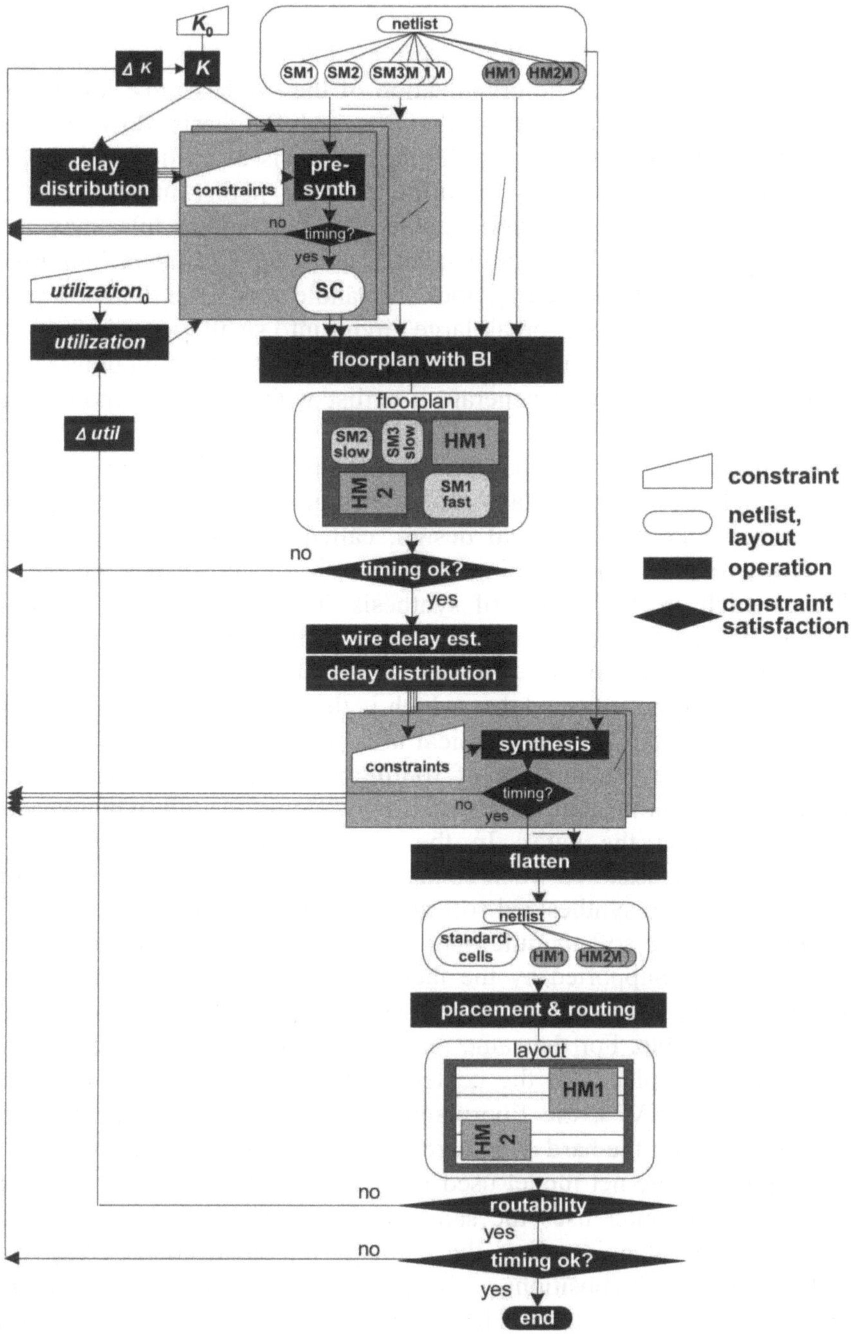

Figure 5.9: The Physical Synthesis flow.

Here it is assumed that all the soft modules will be implemented by standard-cells eventually. The organization of the synthesizable logic in soft modules is not a requirement of the Physical Synthesis flow. Powerful commercial synthesis tools can handle the synthesis of millions of the gates (perhaps these tools internally partition the big netlist into smaller soft modules); so partitioning the synthesizable logic into soft modules may not be necessary. Since the synthesis tool available to us, *SIS* (*Sequential Interactive Synthesis*, developed at Berkeley), does not handle very large flattened netlists well, we use soft modules to break large circuit into small pieces before using *SIS*. The synthesis results (standard-cell netlist) of the soft modules can be merged or flattened later to generate a netlist with only hard modules and standard-cells.

The input to the flow includes a netlist of hard and soft modules as well as a target clock cycle. To obtain more accurate wiring estimation before synthesis, some kind of physical design, called floorplanning, needs to be done. However the floorplanning requires sizes and shapes of the soft modules, which are the results of synthesis. To break this loop, each soft module is pre-synthesized (pre-synth) with estimated arrival and required times on the pins to obtain an estimated size. One important parameter is the area utilization of the soft modules, which is the ratio of the pure gate area to the layout area of the module. Typical area utilizations range from 30% to 90%, depending on the number of routing layers and the interconnection complexity. In addition, buffer insertion may require certain white space be kept when placing the gates. In the Physical Synthesis flow, the area utilization must be assumed when sizing the soft modules. How to determine the shapes of the pre-synthesized soft modules remains a question. In fact the gates belonging to a soft module may not be finally placed within a clearly shaped rectangle. Supported by the fact that most soft modules are smaller than hard modules like RAMs and ROMs, shaping the soft modules into squares is reasonable. For the same reason, the pins of a soft module can be regarded as being placed at the center of the square. As the soft modules become "hard" ("HM"), the floorplanning begins. It is a pure block-level placement, handling the hard and "hard" modules. All interconnections at this stage are global. The net model used in this step is Rectilinear Steiner Tree, because the real router uses the same model. Buffer insertion takes place during the floorplanning for the global interconnections. When the floorplanning ends, the positions of the hard and "hard" modules are obtained, as well as the wiring structures. The estimated wire delays give the arrival and required times for the pins of the soft modules, which are used by the subsequent real synthesis. Based on the estimated physical information, the real synthesis processes the netlist and maps it to gates. The gate netlists of the soft modules are flattened to get one single netlist containing hard modules, buffers and all the gates. The physical design tool works on this netlist and produces the layout. During the physical design, the hard modules maintain their positions as planned by the floorplanning.

Of course at certain steps of the flow, the constraints may be violated or routing violations may occur. Checkpoints are set up in the flow to catch violations and trigger new iterations. The checkpoints are:

(1) During the pre-synthesis of the soft modules, if a module cannot meet the cycle requirement, the flow loops back to the beginning and starts over with an incremented cycle value.

(2) After the floorplanning with buffer insertion, if the cycle requirement cannot be met due to long wire delays, the flow starts over again with an incremented cycle value.

(3) If any soft module fails to meet the cycle requirement and required times during the real synthesis, the flow starts over again with an incremented cycle value.

(4) After placement and routing, if the cycle requirement cannot be met based on the timing analysis (with precise wire delays), the flow loops back to the beginning and increments the clock cycle.

(5) After placement and routing, if routing violations exist, the flow starts over with a smaller area utilization. A smaller loop that goes back to the placement with a lower area utilization is not done, because this changes the whole floorplan which was produced based on the old area utilization.

Unfortunately, the Physical Synthesis flow may still suffer from the following potential problems:

(1) The delay distribution on paths is hard to determine. A path may start from a register in module A, through a wire, and reach a register in module B, as shown in Figure 5.2. The delay must be distributed among these three parts, the internal partial path in module A from the register to the output pin, the wire, and the partial path in module B from the input pin to the register. There is no useful information for making this distribution at this step, even if the wire length can be accurately estimated. In our experiments with the Physical Synthesis flow, we assume that the path delay, with wire delay deducted, is equally distributed to the modules on two sides.

(2) One important parameter is the area utilization of the soft modules, which is the ratio of the pure cell area and the module area. Area utilization is hard to determine at the early stages. Inappropriate values may cause routability problems that occur at the last stage of the flow.

Unfortunately, smaller area utilization may not guarantee easier physical design, since it leads to larger soft modules, which cause longer internal wires inside the modules and longer global wires. This produces tighter timing requirements, which make synthesis produce larger cell netlists (to make the soft modules faster). Hence physical design, wishing to see more white space for easier routing, may receive more cells and thus even less white space.

(3) Pre-synthesis may produce wrong estimation of the module sizes. When the modules are synthesized, their real sizes may invalidate the floorplan and thus some wiring estimates. Module shapes and pin locations may add to the inaccuracy of the floorplanning.

In practice, some small loops can be made in the flow. But such small loops usually require user intervention, making the flow not fully automated.

5.4. The Module-Based Design Flow

5.4.1. Overview

Our Module-Based design flow is illustrated in Figure 5.10. The flow contains two stages: version generation and physical design.

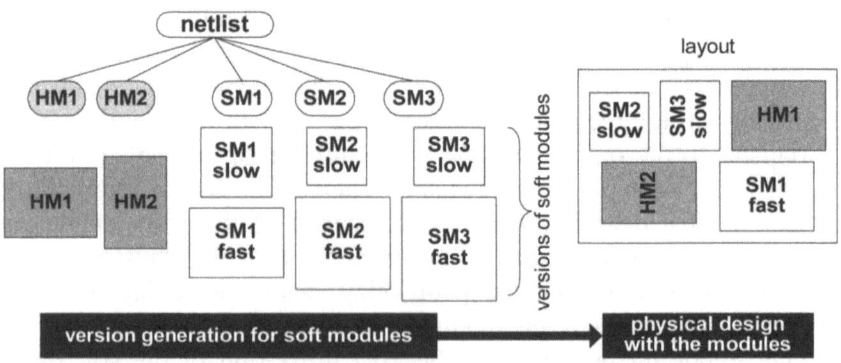

Figure 5.10: Overview of the Module-Based design flow.

The first stage is the preparation of the versions of the soft modules, turning each soft module into a few versions of implementation with different area/delay tradeoffs. Which version is to be used in the final netlist is determined at the second stage, the physical design. The physical design stage uses Fishbone routing with buffer insertion, described in Section 4.3, and in addition, an extra operation, version selection.

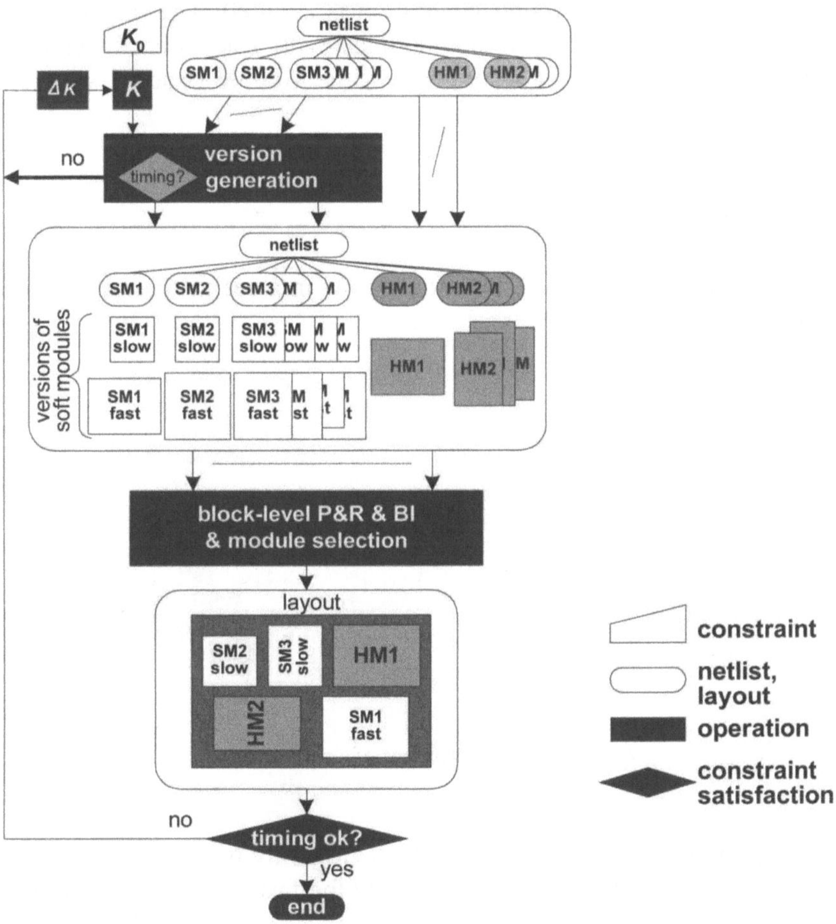

Figure 5.11: The Module-Based design flow.

The Module-Based design flow is shown in Figure 5.11, containing two stages. The first stage is the generation of the versions of the soft modules, turning each soft module into two (or more) versions of hard implementation with different area/delay tradeoffs. Clock cycle may be adjusted due to timing violations. The flow enters the physical design stage when the version generation terminates successfully. The physical design stage uses the Fishbone block-level placement and routing with buffer insertion. The simulated-annealing based Fishbone scheme also makes choices of the versions. Any timing violation after the Fishbone causes a new iteration of the flow with an increased clock cycle.

In Section 5.4.2, the version generation is detailed. The physical design stage, using the integration of the modules with selection of versions, is discussed in 5.4.3. Section 5.4.4 summarizes the Module-Based design flow.

5.4.2. The version of soft module

A version of a soft module is associated with a set of timing constraints, i.e., arrival/required times.

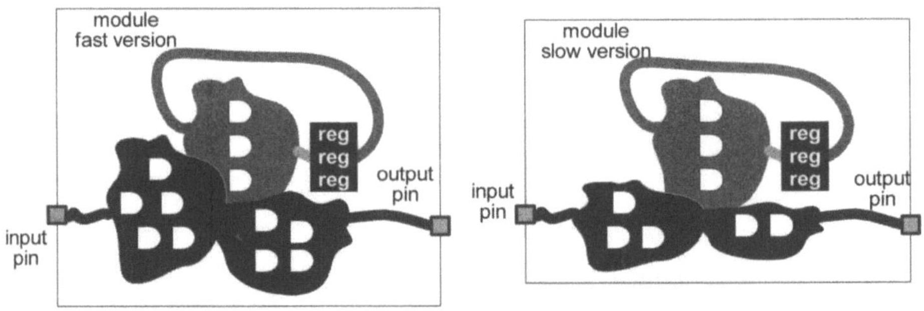

Figure 5.12: The versions of a soft module.

We experimented with a soft module having only two versions: "fast" and "slow", as illustrated in Figure 5.12. Usually the fast version has a larger area than the slow version. The implementation of a version involves both logic synthesis and physical design. After version generation, all modules become hard but with the distinction that the previously soft modules have several versions. Of course, the initial hard modules could have been given in several versions as well, but this was not considered.

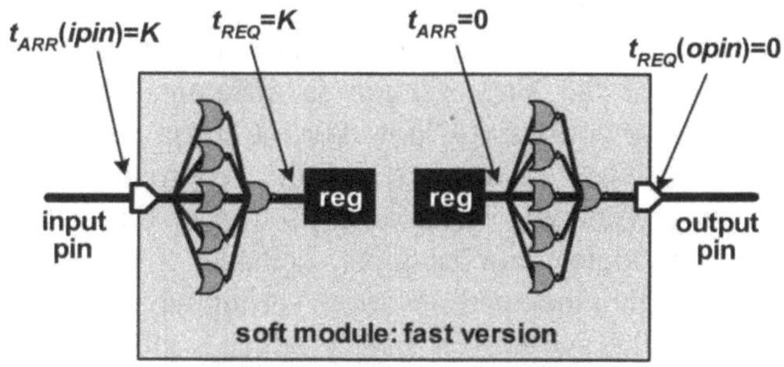

(a) the schematic and constraints

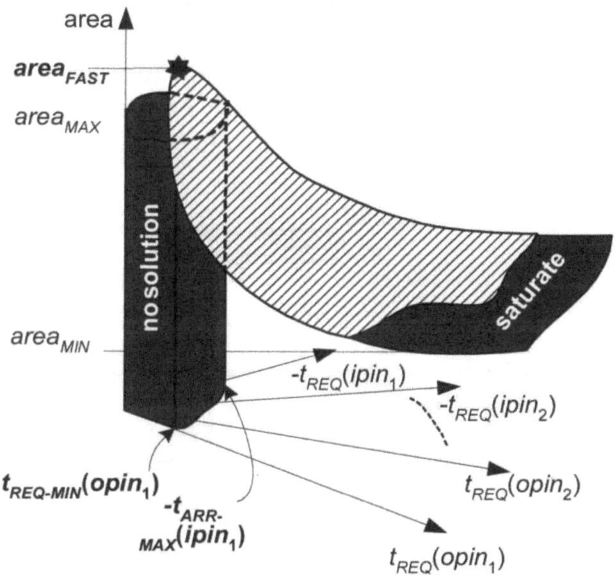

(b) the area/constraints curve

Figure 5.13: The generation of the fast version.

The generation of the fast version is illustrated in Figure 5.13. Let all paths under consideration be single-cycle paths for simplicity. As shown in Figure 5.13(a), the output pin is given the tightest constraint, that is, $t_{REQ}(opin)$=0, and the input pin is constrained by $t_{ARR}(ipin)$=K. Obviously this set of constraints pushes the synthesis into the "no-solution" region of the area/constraint curve. The result is labeled $area_{FAST}$ in Figure 5.13(b). It is fast, in the sense that the inter-module paths from/to this module receive as large delay margins as possible for the wires and other modules. Of course such a version is large in area. The synthesis of the fast version of the soft module does not need any prediction or estimation of wiring effects. A hard module has one single version, and we assume it is a fast version.

The generation of the slow version is different. The slow version is synthesized with loose constraints. However the loose constraints need to be realistic. Arbitrarily loose constraints, infinite required times for instance, may cause the slow version to be totally useless in the following physical integration of the modules. For example, consider a pin of a soft module associated with a single clock path. Without setting any required times on the pins of the module, the synthesizer may produce a slow version that is too slow, the delays of some internal segments of the paths already exceeding one clock cycle. Such slow versions will never be chosen. A reasonable required time set on a pin of the slow version should be some value near the path clock cycle(s) minus the internal delay of the other side (another module) and some wire delay. We will discuss later the choice of the required times for the generation of the soft version.

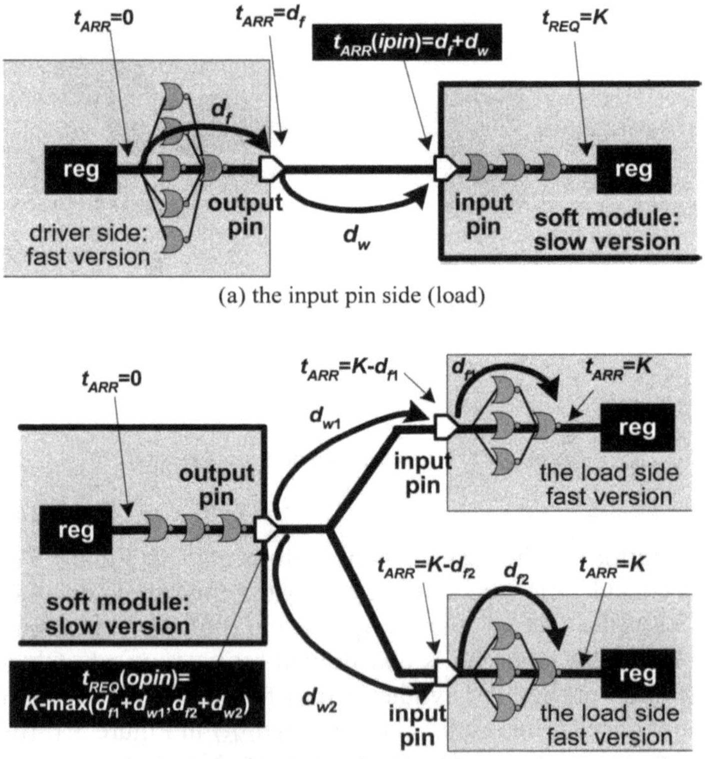

(a) the input pin side (load)

(b) the output pin side (driver)

Figure 5.14: The generation of the slow version.

Figure 5.14 illustrates how slow versions are generated. When the slow version of a soft module is being synthesized, all other modules that this module has connections with are assumed to be fast.

Figure 5.14(a) shows how the arrival time $t_{ARR}(ipin)$ is set on the input pin (load side). The driver side module uses the fast version, which has a known internal delay to its output pin, denoted by d_f. A wire delay d_w, to be clarified later, is assumed. Hence the required time on the input pin can be computed as:

$$t_{ARR}(ipin)=d_f+d_w.$$

The required time of the output pin of the slow version must consider fanout to multiple modules, as illustrated in Figure 5.14(b). The load side modules use fast versions. The two loads have their internal delays, denoted by d_{f1} and d_{f2}. The wire delays of the two pin-pin connections are d_{w1} and d_{w2}. Therefore, the required time on the output pin of the slow soft module is

$$t_{REQ}(opin)=K-\max(d_{f1}+d_{w1},d_{f2}+d_{w2}).$$

From the above discussion, a reasonable sequence of the version generation is to generate fast versions for all soft modules first and then generate the slow versions. Version generation involves both synthesis and physical design, e.g. *SIS* plus ISCPD. Note that the synthesis of the fast versions always results in "no solution". But the synthesis of the slow versions must find a solution; otherwise the slow version will be discarded.

Now the problem is to determine d_W when creating the set of constraints for the slow version. Instead of using techniques like floorplanning to estimate d_W, we conjecture that a module, if using a slow version, is likely to be placed close to the modules it is connected to, reasoning that short wire lengths are the reason for it being able to be slow. This "closeness" is the basis for determining d_W.

An example of deriving the required time for an output pin of a slow soft module (module A) is shown in Figure 5.15. The load side, the fast version of module B, contributes a known internal delay d_{fB} to the path delay. The size of the slow-version of module A is not known yet. But the fast version has been synthesized and laid out, so its area $area_{FAST}(module_A)$ is known. Using the area of the fast version to approximate that of the slow version is safe, because the slow version is usually smaller than its fast version. The slow version is assumed to be a square with area $area_{FAST}(module_A)$. Module B (fast) has been physically designed, so not only its area $area_{FAST}(module_B)$ but also its X and Y sizes are known. We simply use a square with the same area to approximate module B, because otherwise the orientation of the module needs to be guessed if its X and Y sizes are not similar. To make such approximation safe, we set some aspect ratio constraint when doing the physical design for the soft modules. The two square modules are assumed to be placed corner to corner.

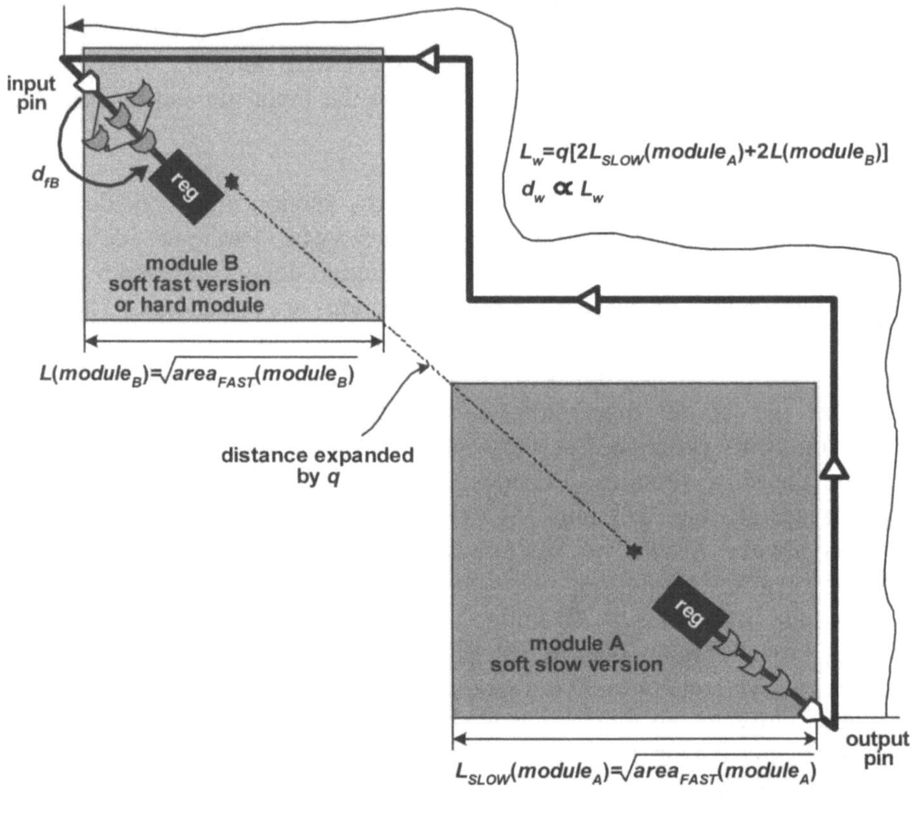

Figure 5.15: Deriving the constraints for the soft slow version.

The "closeness" is measured by the center-center distance of the two modules with an expansion factor of q, as shown in the figure. The Manhattan wire length can now be obtained:

$$\begin{aligned} L_W &= q\left[2L(module_A) + 2L_{SLOW}(module_B)\right] \\ &= 2q\left[\sqrt{area_{FAST}(module_A)} + \sqrt{area_{FAST}(module_B)}\right] \end{aligned}.$$

This assumes that two pins under consideration located on the two extreme corners of the modules. The Manhattan connection with good buffering leads to $d_w \propto L_w$. If the net is multi-fanout, we consider each output-input pin pair with the above method and choose the tightest one to get the required time on the output pin, i.e.,

$$t_{REQ}(opin) = K - \max_{i \in fanout}(d_{fi} + d_{Wi}).$$

The arrival times of the input pins of the slow version can be derived similarly.

The derivation of the arrival/required times does not rely on floorplanning. The design parameter, q, is critical; it is similar to the area utilization parameter used by standard-cell placers. Experimental results show that 1.5~5.0 are feasible values for q. A single q is used for all soft modules. A large q may produce too tight constraints on the slow version, possibly leading to unsuccessful synthesis. We use q=3.0 to run the experiments and report our results. Some pins of the slow version may have positive slacks after synthesis, meaning that modules with connections to the current module through these pins may be placed farther than the distance expressed by q. If two connected modules are placed closer than specified by q, both might use slow versions.

In our experiment, the versions of the soft modules are implemented (both logically and physically) with either standard-cell or Checkerboard. Timing satisfaction is always checked, and obviously inferior implementations are ruled out immediately.

The process of version generation is summarized in Figure 5.16, starting with $K=K_0$, where K_0 is the user-given starting cycle. All the fast standard-cell (SC) versions are generated first. Any timing violation in this step causes an update of the clock cycle, $K=K+\Delta K$, and all previously built fast versions are discarded. Note that the "timing ok?" in this step checks the timing of the internal paths. When all the fast SC versions are generated, an "obvious timing violation" test is performed. This test checks, all inter-module paths with both driver and load sides using fast SC versions, whether the timing constraints are met (assuming small wire delays, computed also based on the "closeness" described by q). If the test fails, there is no way to find a solution for the current clock cycle. If the test is passed, the process enters the next step, the generation of the slow SC versions. The constraints on the pins are created based on the results of the fast versions and the parameter q. Here "timing ok?" tests both internal paths and inter-module paths (slacks on pins) for the module. If the test fails on a particular module, the version is simply discarded rather than launching a new iteration with a modified clock cycle. The main purpose is to save run time. In addition, an "obviously inferior" test is run to discard any slow version, if

$$area_{SLOW} \geq 0.95 arae_{FAST}.$$

Thus, if a slow version is not sufficiently smaller (<0.95) than the fast version, it is discarded also. All modules can be synthesized with circuit structures other than standard-cell. Checkerboard (CB) is also used in our flow. Since the design of Checkerboard structures involves simulated-annealing, its run-time may become significant. Hence the Checkerboard module generation is only done once, after the SC versions are built. The fast CB is produced in a similar way to the fast SC generation. The slow CB version generation borrows the timing constraints from the slow SC. Any timing failure only causes the exclusion of the slow CB module. The "smaller than fast (slow) SC?" test may replace the SC version with the corresponding CB if the latter has a smaller area than the former. An "obviously inferior" test is also run for the slow CB version.

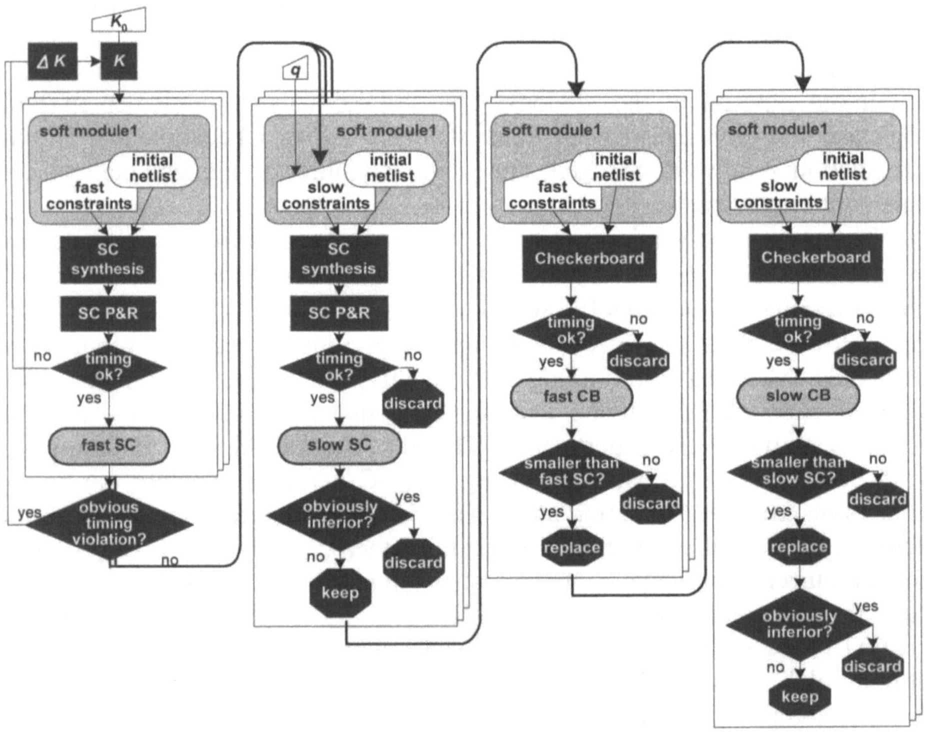

Figure 5.16: The process of the version generation.

When version generation is complete, each soft module has a fast version and possibly a slow version. The versions can be either standard-cell or

Checkerboard. The clock cycle is only adjusted in the fast SC generation step. Since the fast versions of separate modules can be synthesized in parallel, version generation can be sped up using multi-processors. Similarly, the versions of multiple soft modules can be created simultaneously. Note that whenever the clock cycle, K, is changed, all versions are discarded and re-generated using the new clock cycle.

5.4.3. The physical design stage

The task of this stage, besides placement and routing of the modules, is to select versions such that all timing constraints are satisfied and the chip area is reduced. In some sense, the Module-Based design flow postpones the area/delay tradeoffs to the physical design stage. The Fishbone scheme with buffer insertion, described in Section 4.3, is adopted as the physical design algorithm of the Module-Based design flow since selection of a version depends critically on the routing. The only modification is that now a former soft module can have one or two versions. Since the Fishbone algorithm is based on simulated-annealing, randomly making version selection is an easy adaption.

5.4.4. Summary of the Module-Based design flow

Following are several important and interesting questions that are worth discussing:

(1) **Why only two versions, fast and slow?** In fact we can create more versions. This is equivalent to using different q's to generate the non-fast versions of a soft module. But the generation of each version involves synthesis and physical design (SC or CB). If the fast and slow versions of a soft module have similar area/constraints, introducing new intermediate versions may have very little effect on the quality of the final layout, while more versions results in longer run time. The removal of "obviously inferior" slow versions described in the last subsection already takes action to respond to this kind of inefficiency. However, if the fast and slow versions of a soft module have dissimilar area/constraints, some intermediate versions should be created. We have not implemented this in the flow.

(2) **Does the Module-Based flow waste chip area, compared with flattened standard-cell design?** Modular synthesis may produce larger area than flattened synthesis, because the latter can synthesize

across module boundaries. However flattened synthesis also blurs local and global connections, shifting the burden to the physical design. On the other hand, soft modules are only part of the chip, and usually hard modules like RAM and ROM occupy a large portion. Thus the choice of module-based or flattened-cell-based design styles for the same clock cycle will have little effect on the final layout area. Since, clock cycle is a higher priority than chip area, consuming more chip area to get a smaller cycle time is both worthwhile and necessary.

(3) **Is the Module-Based flow similar to the bottom-up design flow?** No. Bottom-up flows do not provide versions for the soft modules. When turning a soft module into a hard module, the bottom-up flow needs to make wiring estimation and solve the delay distribution problem.

(4) **Why do we use regular fabrics like Checkerboard, ISCPD and Fishbone in this flow?** Regularity is a feature, which can provide better guarantees, that the layout designed by a CAD tool is replicated in the fabrication [MO03b]. Manufacturability is another important issue in DSM. The reported results of Checkerboard, ISCPD and Fishbone show that they can compare favorably with standard-cell, a conventional standard-cell physical design algorithm, and a block-level placement and routing algorithm, respectively, in terms of area/delay. Making use of these fabrics is not the reason why the new flow can provide better results than Physical Synthesis which uses standard-cell and a normal physical design algorithm. On the other hand, if manufacturability is a concern, this suggests that the quality of the design might be maintained even if we were to make the whole chip out of these regular fabrics.

5.5. Comparison of the Two Design Flows

5.5.1. Generation of the testing examples

Although industry has provided various benchmark examples to test newly developed algorithms, these examples are limited to logic synthesis and placement [LGSY91, MCNC92, ISPD98]. There are no existing benchmark examples for a design flow. One of the reasons is that such examples may disclose important details of intellectual property products owned by companies.

Thus a diagram-based circuit generator was created. The basic idea is to use circuit diagrams to generate circuit examples. A circuit diagram describes the modular structure of a circuit system and the interconnections of the modules. But the detailed netlist of each module is unknown. Circuit diagrams are easier to obtain than the real circuits. With a diagram, the hard modules such as memory blocks are exposed with respect to the widths of their data and address buses. Since there are no details for the soft modules, which implement random logic like control machines, we randomly associate existing synthesis benchmarks with them.

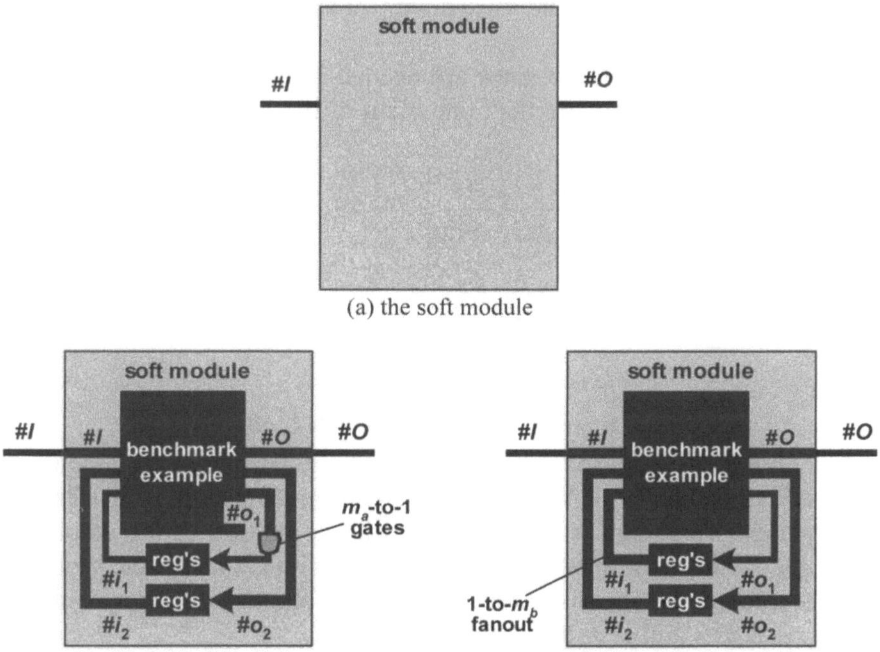

(a) the soft module

(b) $(\#I_B-\#I)<(\#O_B-\#O)$

(c) $(\#I_B-\#I)\geq(\#O_B-\#O)$

Figure 5.17: The pin matching.

Pin matching must be obeyed when the association is established, as shown in Figure 5.17. Given a soft module with $\#I$ input pins and $\#O$ output pins, we have to find from a pool of benchmark examples a circuit that has very close pin counts. We can always find a benchmark circuit with $\#I_B$ input pins and $\#O_B$ output pins, where $\#I_B>\#I$ and $\#O_B>\#O$. Closeness is measured by $(\#I_B-\#I)+(\#O_B-\#O)$. Excessive pins are handled as follows:

(1) $(\#I_B-\#I)<(\#O_B-\#O)$. Feedback loops are created to connect the excessive outputs and inputs, as illustrated in Figure 5.17(b):

$$m_a = \left\lceil \frac{\#O_B - \#O}{\#I_B - \#I} \right\rceil$$

$$\#i_1 = \left\lfloor \frac{\#O_B - \#O}{m_a} \right\rfloor$$

$$\#o_1 = m_a \times \#i_1$$

$$\#i_2 = \#o_2 = \#I_B - \#I - \#i_1 = \#O_B - \#O - \#o_1$$

In the equations, m_a is an integer ($m_a \geq 2$) scaling factor that reduces $\#o_1$ to $\#i_1$.

(2) $(\#I_B-\#I)\geq(\#O_B-\#O)$. No gates are needed in this case. But $\#i_1$ registers may need to drive more than one input of the benchmark circuit with a scaling factor of m_b:

$$m_b = \left\lceil \frac{\#I_B - \#I}{\#O_B - \#O} \right\rceil$$

$$\#o_1 = \left\lfloor \frac{\#I_B - \#I}{m_b} \right\rfloor$$

$$\#i_1 = m_b \times \#o_1$$

$$\#i_2 = \#o_2 = \#I_B - \#I - \#i_1 = \#O_B - \#O - \#o_1$$

In a circuit system, especially an SOC, a large portion of the modules are hard, such as IP blocks, RAM and ROM. The random logic is distributed in soft modules, which can occupy an insignificant portion of the whole chip. Therefore, if the hard modules like the IP and parameterized blocks are correctly created, the circuit generated with this approach can approximate the real chip size and global interconnection structure.

5.5.2. The 0.18-micron technology

In our experiments, a 0.18-micron technology was used. Some important design parameters are given in Table 5.1. Five metal layers are available, the three lower layers for short or local connections and the two upper layers for long or global connections.

Table 5.1: The technology information

parameter		unit	value
gate height		μm	6.0
buffer area		μm^2	15.0
buffer intrinsic delay		ps	11.0
buffer load dependent delay		ps/pF	439.0
buffer input capacitance		pF	0.0059
routing pitch	layer1,2	μm	0.6
	layer3	μm	0.7
	layer4,5	μm	0.8
sheet resistance	layer1-4	Ω/□	8.0e-2
	layer5	Ω/□	7.0e-2
area capacitance	layer1,2	$pF/\mu m^2$	3.9e-7
	layer3	$pF/\mu m^2$	4.0e-7
	layer4,5	$pF/\mu m^2$	3.7e-7
fringe capacitance	layer1	pF/μm	5.1e-6
	layer2	pF/μm	4.5e-6
	layer3	pF/μm	5.3e-6
	layer4	pF/μm	5.7e-6
	layer5	pF/μm	6.5e-6

5.5.3. The testing circuits

Five circuit templates were created, covering several types of applications like complex processors, wireless computing chips and real time image processing chips. A pool of 77 circuits from the MCNC synthesis benchmark suite [MCNC92] was built for the generation of the soft modules.

The benchmark examples for the design flow were generated from the templates by setting the bus widths, memory sizes and other parameters that affect the chip size and interconnection complexity. The target (initial) clock cycle of a design took the cycle of the slowest hard module in the design (memory blocks and other hard modules may have different smallest working cycles). The characteristics of the examples are given in Table 5.2.

Table 5.2: The testing circuits

circuit	application	#HM	#SM	#I/O	#global net	target cycle (ps)	comment
A1	CISC	11	8	114	668	2000	
A2	CISC	12	8	174	964	2000	
A3	unknown application	10	10	166	1160	2000	
A4	wireless computing	17	41	403	2832	2222	many soft modules
A5	real time image processing	10	15	158	1008	5000	large hard modules, large chip

5.5.4. The comparison

Logic synthesis used *SIS*. The physical design of the Physical Synthesis flow used the *Cadence Silicon Ensemble* version 5.3 (*QPlace* plus *Warp Router*) coupled with a high-quality floorplanner that we wrote based on simulated-annealing (Steiner Tree and buffer insertion) using methods described in [MURA96, VANG90]. This floorplanner with buffer planning was used in Section 4.3 where it was shown that it can offer slightly better area/delay than Fishbone-B, if these two are treated as standard-alone algorithms. The Fishbone scheme requires module pins be placed on certain routing grids. Slight modifications were made for the pin locations of the hard modules, but this should not add to the difficulty of the routing by *Warp Router*. The SC and CB modules created in the version generation of our flow took into account the pin location requirement. No routing for power, ground and clock nets were done in the experiments. These nets are normally be routed by special routers on reserved metal layers.

When iterations occur, the increment of the clock cycle was chosen as:

$$\max\{-MostNegativeSlack + 20\text{ps}, 100\text{ps}\}.$$

Tables 5.3 and 5.4 record the runs of the two flows, respectively. Each row in the tables is one iteration of the flow. The bold items in the table indicate where timing or routing violations invoked new iterations.

The area/cycle results of the five examples are compared in Figure 5.18. On average, the Module-Based flow gives 13% smaller cycle times than the Physical Synthesis flow. Chip areas are not directly comparable, since the chips produced by the two flows may work under different clock periods, and the reduction of the clock cycle had higher priority than the area in the experiments. The Module-Based flow, on average, takes 77% less run time than the Physical Synthesis flow. Although precisely testing memory usage is hard, we have observed that the Warp Router used 1.6GB memory when routing A5, while the maximum memory used by the Module-Based flow was below 180MB.

Table 5.3: The record of the Physical Synthesis flow
A: area (μm^2); Cycle: (ps) S: most-negative-slack (ps); V: routing_violation; T: run time (s)

(part 1 of 2)

circuit	cycle	stage: pre-synthesis	stage: floorplan with buffer insertion	stage: synthesis + flatten	stage: place + route	total time
A1	2000	**S=-4**; T=9				
	2224	**S=-43**; T=39				
	2487	**S=-58**; T=139				
	2587	S=0; T=75	A=8.92e5; S=0; T=1789	A=8.92e5; **S=-31**; T=6		
	2687	S=0; T=75	A=9.02e5; S=0; T=1770	A=9.02e5; **S=-11**; T=43		
	2918	S=0; T=75	A=9.15e5; S=0; T=1731	A=9.15e5; **S=-17**; T=42		
	3018	S=0; T=75	A=9.17e5; S=0; T=1696	A=9.17e5; S=0; T=144	A=9.17e5; S=0; V=0; T=209	2.2h
A2	2000	**S=-113**; T=81				
	2133	**S=-24**; T=32				
	2233	**S=-62**; T=81				
	2333	S=0; T=47	A=9.07e5; S=0; T=2152	A=9.07e5; **S=-153**; T=90		
	2506	S=0; T=49	A=8.95e5; S=0; T=1992	A=8.95e5; **S=-7**; T=89		
	2606	S=0; T=47	A=8.99e5; S=0; T=2088	A=8.99e5; **S=-106**; T=86		
	2732	S=0; T=47	A=8.99e5; S=0; T=2120	A=8.99e5; S=0; T=87	A=8.99e5; S=0; **V=8**; T=671	
	2732	S=0; T=49	A=9.20e5; S=0; T=2096	A=9.20e5; S=0; T=86	A=9.20e5; S=0; V=0; T=196	3.4h

(part 2 of 2)

circuit	cycle	stage: pre-synthesis	stage: floorplan with buffer insertion	stage: synthesis + flatten	stage: place + route	total time
A3	2000	**S=-327**; T=178				
	2347	S=0; T=174	A=1.08e6; S=0; T=2272	A=1.08e6; **S=-340**; T=201		
	2707	C=0; T=65	A=1.10e6; S=0; T=2240	A=1.10e6; **S=-37**; T=204		
	2807	S=0; T=65	A=1.09e6; S=0; T=2264	A=1.09e6; S=0; T=171	A=1.09e6; S=0; **V=2**; T=569	
	2807	S=0; T=67	A=1.11e6; S=0; T=2240	A=1.11e6; S=0; T=169	A=1.11e6; S=0; V=0; T=188	3.0h
A4	2222	**S=-105**; T=177				
	2347	**S=-145**; T=318				
	2512	S=0; T=311	A=1.76e7; S=0; T=7912	A=1.76e7; **S=-208**; T=195		
	2740	S=0; T=284	A=1.76e7; S=0; T=7864	A=1.76e7; **S=-15**; T=195		
	2840	S=0; T=285	A=1.76e7; S=0; T=7896	A=1.76e7; **S=-143**; T=206		
	2963	S=0; T=284	A=1.81e7; S=0; T=7928	A=1.81e7; S=0; T=306	A=1.81e7; S=0; **V=6**; T=7115	
	2963	S=0; T=281	A=1.77e7; S=0; T=7808	A=1.77e7; S=0; T=302	A=1.77e7; S=0; **V=16**; T=8366	
	2963	S=0; T=308	A=1.80e7; S=0; T=8024	A=1.80e7; S=0; T=303	A=1.80e7; S=0; **V=11**; T=7277	
	2963	S=0; T=282	A=1.85e7; S=0; T=7840	A=1.85e7; S=0; T=302	A=1.85e7; S=0; **V=2**; T=7390	
	2963	S=0; T=283	A=1.86e7; S=0; T=7880	A=1.86e7; S=0; T=307	A=1.86e7; S=0; **V=9**; T=7496	
	3083	S=0; T=285	A=1.77e7; S=0; T=7768	A=1.77e7; S=0; T=272	A=1.77e7; S=0; **V=12**; T=7090	
	3083	S=0; T=283	A=1.87e7; S=0; T=7816	A=1.87e7; S=0; T=269	A=1.87e7; S=0; V=0; T=7477	38h
A5	5000	S=0; T=154	A=2.52e8; S=0; T=40392	A=2.52e8; S=0; T=148	A=2.52e8; S=0; V=0; T=23472	18h

Table 5.4: The record of the Module-Based flow
A: area (μm^2); C: cycle (ps); V: routing_violation; T: run time (s)

circuit	cycle	stage: module synthesis & P&R	stage: chip P&R with buffer insertion	total time
A1	2000	**S=-204**; T=12		
	2224	**S=-243**; T=348		
	2487	**S=-58**; T=411		
	2587	S=0; T=542	A=9.08e5; S=0; V=0; T=958	0.62h
A2	2000	**S=-113**; T=61		
	2133	**S=-24**; T=83		
	2233	**S=-62**; T=277		
	2333	S=0; T=409	A=1.67e6; S=0; V=0; T=1390	0.61h
A3	2000	**S=-327**; T=217		
	2347	S=0; T=842	A=1.72e6; S=0; V=0; T=1674	0.75h
A4	2222	**S=-105**; T=281		
	2347	**S=-145**; T=1330		
	2512	S=0; T=2892	A=2.25e7; S=0; V=0; T=5445	2.8h
A5	5000	S=0; T=2415	A=2.51e8; S=0; V=0; T=20149	6.3h

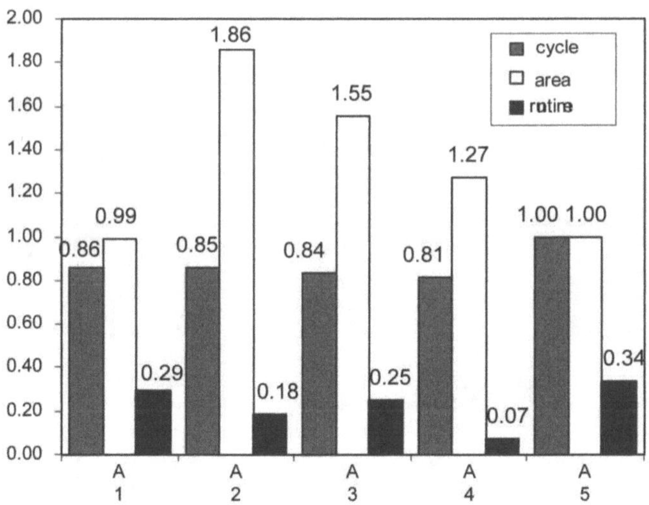

Figure 5.18: Comparison of the two flows
(Module-Based/Physical Synthesis)

In the Physical Synthesis flow, the iteration from the pre-synthesis takes little time. The floorplanning always took significant run time, so any iteration triggered after that is expensive. The physical design for large circuits was slow. Routability problems did occur in the flow (A3 and A4). Furthermore, example A4 experienced the area utilization problem mentioned in Section 5.4.4 (potential problem (3)). When routing violations occurred at C=2963, the area utilization was adjusted several times in the hope that more white space may be generated to ease the routing difficulty. But this did not seem to help; and note that such attempts were very timing consuming because the iterations involved the biggest loop in the flow. In the experiment, we simply added an upper limit for the number of such attempts and switched to a new clock cycle after this.

In the Module-Based design flow, no iteration of the Fishbone place and route ever happened. Iterations happened only during the version generation stage. Routability problems never occurred.

5.5.5. Case study: A1

The system diagram of A1 is illustrated in Figure 5.19.

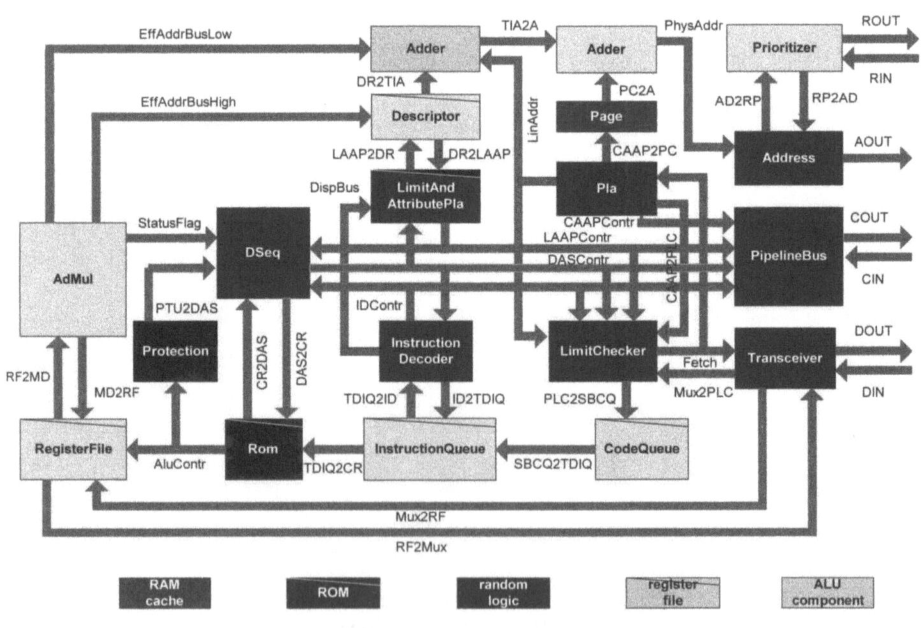

Figure 5.19: The system diagram of A1

There are five types of modules: RAM or cache, ROM, random logic, register file and ALU type, among which all but the random logic type are hard modules. The modules of the random logic category are soft, and each is associated with a *blif* (Berkeley Logic Interchange Format) file, as described in 5.5.1. The size of the circuit is mainly controlled by data and address bus sizes, because these determine the sizes of the RAM, ROM, register files and ALU modules. In this example, the data buses such as TIA2A are 32-bit wide. Address buses are of different widths, ranging from 9 to 15 bits.

The versions of soft modules generated for A1 in the Module-Based design flow are listed in Table 5.5. The selection made in the physical design stage is also shown. Note that the first and last modules and their slow versions discarded, because the versions were "obviously inferior".

Table 5.5: The selection of the versions for A1

soft module	area (μm^2) fast		slow		selection
ProtectionTestUnit	59.2×78.4	(SC)	discarded	(SC)	fast
DecodeAndSequencing	155.2×174.4	(SC)	126.4×174.4	(SC)	slow
InstructionDecoder	188.8×107.2	(CB)	184.0×97.6	(SC)	slow
ControlAndAttributePla	251.2×107.2	(SC)	251.2×97.6	(SC)	fast
Prefetcher	169.6×265.6	(CB)	174.4×246.4	(SC)	slow
AddressDriver	270.4×236.8	(SC)	246.4×193.6	(CB)	slow
PipelineBusSizeControl	155.2×179.2	(SC)	131.2×188.8	(SC)	slow
MuxTransceiver	160.0×150.4	(CB)	discarded	(SC)	fast

The layout picture of the chip produced by the Module-Based design flow is shown in Figure 5.20(a). The layout pictures of a Checkerboard module and a standard-cell module are also shown. The layout picture of the chip produced by the Physical Synthesis flow is shown in Figure 5.20(b).

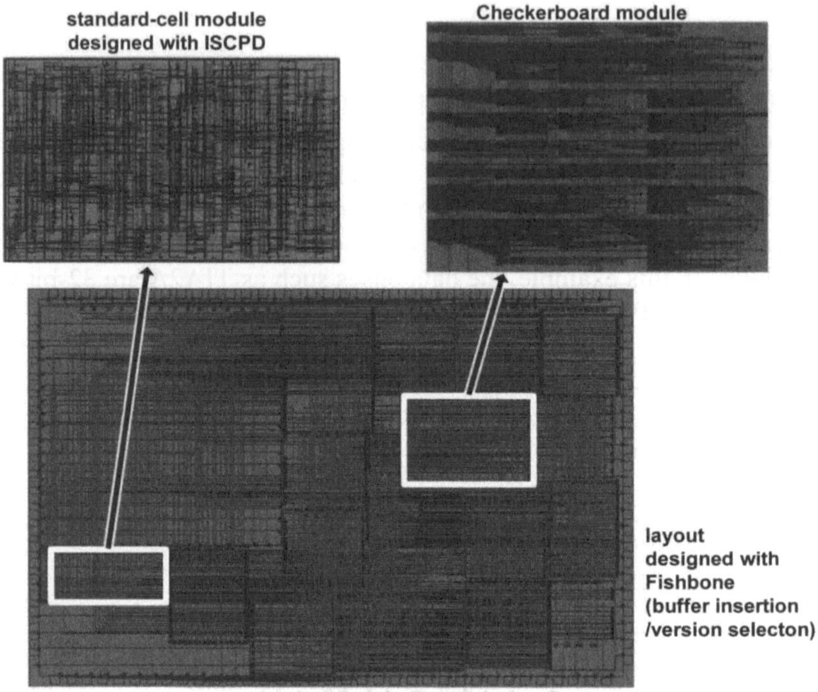

(a) the Module-Based design flow

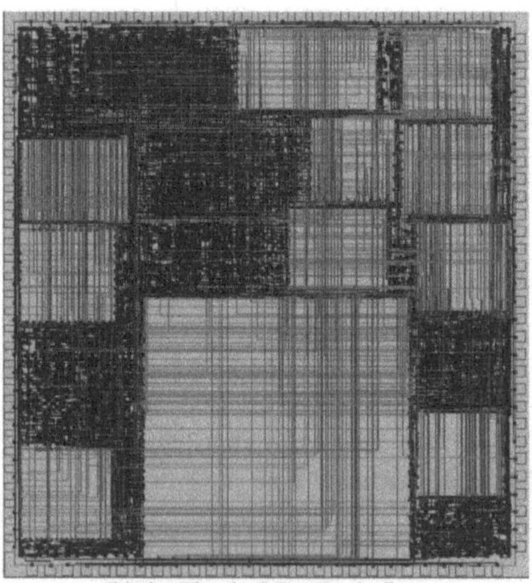

(b) the Physical Synthesis flow

Figure 5.20: The layout pictures of A1

5.6. Summary

The Module-Based design flow was presented, which contains a version generation stage and a physical design stage. Key elements are the creation of versions of the soft modules and the use of a physical design stage that does simultaneous place and route using accurate wire delays. A soft module is implemented in two versions, fast and slow. The version generation does not rely on wiring estimation, but selects the versions at the physical design stage to meet real wiring situations. New structures like Checkerboard and new design methodologies like ISCPD are both adopted in the module generation stage. The Fishbone scheme with buffer insertion is adopted in the physical design stage, modified by including version selection. Experimental results show that the Module-Based design flow achieves shorter clock periods than the Physical Synthesis approach and uses less time.

Chapter 6

CONCLUSION

IC design has begun to benefit from the Deep Sub-Micron (DSM) technology, which offers higher density of transistor integration and higher performance. However, DSM technology has introduced many new problems, challenging the feasibility and efficiency of current CAD tools. One of the most critical issues is timing closure, which is mainly due to the growing importance of the wiring effects. The separation of logic synthesis and physical design, two sequential stages in a traditional design flow, has become inadequate since these two stages are tightly coupled by wiring effects. Timing convergence can no longer be easily guaranteed even if iteration between the two stages is done.

Another critical problem is manufacturability, which not only directly affects the yield of the IC products, but also the predictability of the behavior of the manufactured product, including timing. Research on Design-for-Manufacturability (DFM) is still at its very early stages; there are no quantitative criteria for the quality of the layout generated by CAD tools. But it is clear that regular layout structures can greatly simplify the work of analyzing and adjusting layout patterns to achieve better manufacturability.

Manufacturability can be pursued through the development of regular circuit and interconnection structures. Several PLA-type regular circuit structures were presented in this book, including River-PLA, Whirlpool-PLA and Checkerboard. Design methodologies for these structures were also developed. An integrated placement and routing algorithm for the widely used standard-cell structure was also implemented. The "Integrated Standard-Cell Physical Design" (ISCPD) algorithm employs a simple regular spine net topology, which enables fast and predictable wiring construction. The algorithm automatically adjusts area utilization to guarantee routability. The above four structures can be used to create regular modules. Other types of regular modules could be used as well. Some of these are discussed at the end of this chapter.

To inter-connect the regular modules with a regular wiring scheme, a block-level placement and routing scheme called Fishbone was developed and presented also. This grid-routing system reduces the routing problem to the assignment of wires to the columns and rows, and then doing simple left-edge routing in the columns and rows. The notion of the base/virtual pin pair was used to significantly reduce the risk of meeting routability problems. The fast wiring construction of Fishbone allows an integration of the block placement and routing in a simulated-annealing framework, since the time for routing is a small fraction of the computation. Real wire delays now appear in the cost function. A buffer insertion feature was also integrated into Fishbone for accurate timing during global placement and routing.

These regular circuit and interconnection structures produce simple layout patterns, from which the mask engineering work for manufacturability can benefit. For example, the number of vias is essentially minimized. These structures also offer better area/delay predictability, which are potentially suitable for timing driven design flows.

To tackle the timing closure problem at the global level, a Module-Based design flow was presented and implemented. There are two basic elements of this flow. One is to build multiple versions for each module with different area/delay tradeoffs. This allows the physical design stage to choose an appropriate version to meet the timing requirements. The second element is to use the Fishbone scheme with buffer insertion for the physical design stage. This allows for rapid routing and accurate delay estimation during placement, so that the appropriate version choice can be made at each step of the placement algorithm. Although we cannot claim that the timing closure problem is completely solved with the use of the Module-Based design flow, when compared directly to Physical Synthesis, our experiments, indicate that the Module-Based flow does produce better achievable clock cycles and shorter run-times. One benefit of the latter is that users can get feedback quickly and if necessary, make design changes without waiting a long time to get a failure notice.

The following list summarizes the contents of the book:

(1) ISCPD: An integrated standard-cell physical design algorithm. It guarantees routability and removes the requirement for the user to supply a guess about area utilization. The area and delay achieved by ISCPD are comparable to those of a commercial tool using separate placement and routing, while ISCPD's run time is only 37% of the commercial tool.

(2) River PLA: A regular circuit structure and its synthesis algorithm. RPLA is a linear stack of multiple-output PLAs, where adjacent PLAs are connected via river routing. Thus only two metal layers are needed. Experiments showed that the areas of RPLAs are on average slightly larger than SC while the delays of these two structures are similar. However, RPLAs provides higher regularity and their delays are predictable with high accuracy. A via-programmable version of the RPLA, called Glacier PLA, where only the masks of the vias are used to customize the logic was developed also.

(3) Whirlpool PLA: A four-level regular structure and its synthesis algorithm. A WPLA is a cyclic chain of four programmable NOR arrays. The compact layout of a WPLA (about half that of SC area) is a result of its unique structure as well as a special four-level logic synthesis algorithm called *Doppio-ESPRESSO*. Although experiments showed that these are applicable to smaller designs of less than 1K gates, due to their limitation of 4 levels of logic, within this range, they were shown to be superior to SC in both area and delay.

(4) Checkerboard: A regular array structure and its synthesis algorithm. Checkerboard is an array of programmable NOR arrays. They are PLA-like in the sense that two such NOR arrays could make up one PLA. The interconnections between the NOR arrays use a simple predictable spine topology. The logic synthesis of a Checkerboard implementation involves a decomposition of the Boolean network into a network of OR gates and a simulated-annealing based physical design aspect. The regularity of the Checkerboard is very high due to the regularity of the NOR arrays and the regular spine topology routing structure; areas and delays are comparable to those of SC. It also has a via-programmable version.

(5) Fishbone: An integrated block-level placement and routing algorithm. The spine topology used in Fishbone allows quick construction of the interconnections and accurate delay computation based on actual wiring. Routability is enhanced by the use of base/virtual pin pairs, where base pins are made wide at the lower level, whereas the virtual pin can be any location on an upper level which is over the base pin. Compared with a Steiner Tree based algorithm, Fishbone provides similar area and delay, but its run time is 577% shorter. Simultaneous buffer insertion was also implemented in Fishbone for use with global chip-level routing. While the use of the simple spine topology may seem at first a waste of resources, our experiments showed that when combined with placement, the latter adjusts to accommodate the spine topology without much loss of routing resources. At the same time, the base/virtual pin notion alleviates local routing congestion.

(6) The "Module-Based" chip design flow. This represents one attempt to put together all the above elements into a chip-level design flow. One key of the Module-Based chip design flow is that different versions of soft modules are used. These transmit flexibility created early in the flow for use later at the physical design stage. This design flow reduces the possibility of triggering large iterations in the design process by extensively using regular and predictable circuit and interconnection structures. Experimental results showed that the Module-Based chip design flow achieves 13% smaller cycle times than a Physical Synthesis approach, while its run time is 77% shorter.

Other Regular Structures

This book has documented the use of a number of regular structures and put these into a regular design flow. Of course, there are many other regular structures to be invented or already existing. Among the latter are gate arrays, [SMIT97], Path Programmable Logic [GU89], and Via Patterned Gate Arrays (VPGA) [PATE03]. Below are short discussions of each and the different tradeoffs they represent.

In a gate array structure, masks for transistor layers are pre-designed. To implement gates, adjacent transistors are locally connected. The allowed gate positions, usually called sites, are not totally free, because of the fixed transistor locations. Design automation tools for gate arrays include (1) logic synthesis which produces the gate netlist, (2) placement of gates on the sites and (3) routing. Hence gate array designs are similar to standard-cell designs, except that the placement of gates must obey site constraints. The loss of this design freedom is compensated by quick time-to-market and low unit cost.

Path Programmable Logic (PPL) circuits are to some extent similar to gate arrays, in which transistors are laid out in an array and gates are implemented in pre-defined "sites" by connecting a group of transistors. Initially a "sea-of-wires" covers the whole layout area. Whenever a gate (a group of transistors) is placed, the wires over that site are pruned, while wires around the site are left intact. On the other hand, the PPL structure is similar to PLA; the signal flows in a PPL circuit turn 90° as they pass one level of transistors. The authors of PPL claimed that PPL outperformed gate arrays in terms of area [GU89].

The building block of Via Patterned Gate Arrays (VPGA) is a Look-Up Table (LUT). Logically a VPGA is similar to a Field Programmable Gate Array (FPGA), which also uses LUTs as the basic function blocks. However the configuration of the LUTs in a VPGA is implemented through via masks instead of static RAMs as in FPGAs. This greatly reduces the chip size of an VPGA, also making it potentially faster than a FPGA. In terms of interconnection structures, the authors of VPGA [PATE03] discussed crossbar and switch block. Both are implemented with simple and regular metal patterns and configured through via layers. VPGA can be an attractive structure, because of its low mask cost and superiority to FPGA in area and performance.

BIBLIOGRAPHY

[ADYA02] S.N. Adya and I.L. Markov, "Consistent Placement of Macro-Blocks Using Floorplanning and Standard-Cell Placement", *International. Symposium on Physical Design*, Apr 2002, pages 12-17.

[BEDN02] T.R. Bednar, P.H. Buffet, R.J. Darden, S.W. Gould and P.S. Zuchowski, "Issues and Strategies for the Physical Design of System-On-a-Chip", *IBM Journal of Research and Development*, vol. 46, no.6, Nov 2002, pages 661-674.

[BODA01] S. Bodapati and F.N. Najm, "Prelayout Estimation of Individual Wire Lengths", *IEEE Transactions on Computer-Aided Design of Integrated Circuits and Systems*, vol. 9, no. 6, Dec 2001, pages 943-958.

[BORA94] M. Borah, R.M. Owens and M.J. Irwin, "An Edge-Based Heuristic for Steiner Routing", *IEEE Transactions on Computer-Aided Design of Integrated Circuits and Systems*, vol. 13, no.12, Dec 1994, pages 1563-1568.

[BRAY84] R. Brayton, G. Hachtel, C. McMullen and A. Sangiovanni-Vincentelli, *Logic minimization algorithms for VLSI synthesis*, Kluwer Academic Publishers, 1984.

[BRAY90] R. Brayton, G. Hachtel and A. Sangiovanni-Vincentelli, "Multi-level logic synthesis", *Proceedings of the IEEE*, vol. 78, Feb 1990.

[BRYA01] R. Bryant, K-T. Cheng, A. Kahng, K. Keutzer, W. Maly, R. Newton, L. Pileggi, J. Rabaey, A. Sangiovanni-Vincentelli, "Limitations and challenges of computer-aided design technology for CMOS VLSI", *Proceedings of the IEEE*, vol. 89, issue 3, Mar 2001, pages 341-365.

[BURS86] M. Burstein, *Layout Design and Verification*, chapter 4, Elsevier Science Publishers, 1986.

[CADE] http://www.cadence.com/products/pks.html

[CALD00] A.E. Caldwell, A.B. Kahng and I.L. Markov, "Can Recursive Bisection Alone Produce Routable Placement?", *Design Automation Conference*, 2000, pages 477-482.

[CLEI99] D. Clein, *CMOS IC Layout: Concepts, Methodologies and Tools*, Newnes, 1999.

[CONG97] J. Cong, A. Kahng and K. Leung, "Efficient Heuristics for the Minimum Shortest Path Steiner Arborescence Problem with Applications to VLSI Physical Design", *International Symposium on Physical Design*, 1997, pages 88-95.

[CONG00] J. Cong, H. Huang and X. Yuan, "Technology Mapping for k/m-macrocell Based FPGAs", *ACM/SIGDA International Symposium on Field Programmable Gate Arrays*, Feb 2000, pages 51-59.

[CONG01a] J. Cong, "An Interconnect-Centric Design Flow for Nanometer Technologies", *Proceedings of the IEEE*, vol. 89, no. 4, Apr 2001, pages 505-528.

[CONG01b] J. Cong, T. Kong and Z. Pan, "Buffer Block Planning for Interconnect Planning and Prediction", *IEEE Transactions on Very Large Scale Integration (VLSI) Systems*, vol. 9, no. 6, Dec 2001, pages 929-937.

[DHON92] Y.B. Dhong and C.P. Tsang, "High Speed CMOS POS PLA using Predischarged OR Array and Charge Sharing AND Array", *IEEE Transactions on Circuits and Systems II: Analog and Digital Signal Processing*, vol. 39, no. 8, Aug 1992, pages 557-564.

[DUTT95] S.Dutta, S.S.Mahant and S.L.Lusky, "A Comprehensive Delay Model for CMOS Inverters", *IEEE Journal of Solid-State Circuits*, vol.30, No.8, Aug 1995, pages 864-871.

[EISE98] H. Eisenmann and F.M. Johannes, "Generic Global Placement and Floorplanning", *Design Automation Conference*, 1998, pages 269-274.

[ELMO48] W.C. Elmore, "The Transient Response of Damped Linear Networks with Particular Regard to Wideband Amplifiers", *Journal of Applied Physics*, Jan 1948, pages 19:55-63.

[FIDU82] C.M. Fiduccia and R.M. Mattheyses, "A Linear-Time Heuristic for Improving Network Partitions", *Design Automation Conference*, 1982, pages 171-181.

[GANL97] J.L.Ganley, "Accuracy and Fidelity of Fast Net Length Estimates", *ACM VLSI Integration, the VLSI Journal*, vol. 23, no. 2, Nov 1997, pages 151-155.

[GROE89] P. Groeneveld, "Wire Ordering for Detailed Routing", *IEEE Design & Test of Computers*, vol. 6, 1989, pages 6 -17.

[HASH71] A. Hashimoto and J. Stevens, "Wire Routing by Optimizing Channel Assignment within Large Apertures", *Proceedings of Design Automation Workshop*, 1971, pages 155-169.

[HODG03] D.A. Hodges, R.A. Saleh and H.G. Jackson, *Analysis and Design of Digital Integrated Circuits*, McGraw-Hill, 2003.

[HONG00] X. Hong, G. Huang, Y. Cai, J. Gu, S. Dong, C-K. Cheng and J. Gu, "Corner Block List: An Effective and Efficient Topological Representation of Non-Slicing Floorplan", *International Conference on Computer Aided Design*, Nov 2000, pages 8-12.

[HU02] J. Hu, C.J. Alpert, S.T. Quay and G. Gandham, "Buffer Insertion with Adaptive Blockage Avoidance", *International Symposium on Physical Design*, 2002, pages 92-97.

[HU03] B. Hu, Y. Watanabe, A. Kondratyev and M. Marek-Sadowska, "Gain-based technology mapping for discrete-size cell libraries", *Design Automation Conference*, Jun 2003, pages 574-579.

[IGUC01] Y. Iguchi, T.Sasao and M. Matsuura, "Realization of multiple-output functions by reconfigurable cascades", *International Conference on Computer Aided Design*, Nov 2001, pages 388-393.

[ISPD98] The ISPD 98 circuit benchmark suite, http://vlsicad.cs.ucla.edu/~cheese/ispd98.html

[JACK90] M. Jackson, A. Srinivasan and E.S. Kuh, "Clock Routing for High-Performance ICs", *Design Automation Conference*, 1990, pages 573-579.

[KAHN92] A. B. Kahng and G. Robins, "A New Class of Iterative Steiner Tree Heuristics with Good Performance", *IEEE Transactions on Computer-Aided-Design of Integrated-Circuits and Systems*, vol. 11, no. 7, Jul 1992, pages 893-902.

[KAHN01] A.B. Kahng, S. Mantik and D. Stroobandt, "Toward Accurate Models of Achievable Routing", *IEEE Transactions on Computer-Aided Design of Integrated Circuits and Systems*, vol. 20, no. 5, May 2001, pages 648-659.

[KANG95] S.M. Kang, Y. Leblebici and Y. Leblebici, *CMOS Digital Integrated Circuits: Analysis and Design*, McGraw Hill, 1995.

[KAST00] R. Kastner, E. Bozogzadeh and M. Sarrafzadeh, "Predictable Routing", *International Conference on Computer Aided Design*, Nov 2000, pages 110-113.

[KHAT00] S. Khatri, *Cross-talk Noise Immune VLSI Design using Regular Layout Fabrics*, PhD thesis, University of California Berkeley, Electronics Research Laboratory, College of Enginnering, University of California, Berkeley, CA94720, Spring 2000.

[KLEI91] J.M. Kleinhans, G. Sigl, F.M. Johannes and K.J. Antreich, "GORDIAN: VLSI Placement by Quadratic Programming and Slicing Optimization", *IEEE Transactions on Computer-Aided Design of Integrated Circuits and Systems*, no. 10, 1991, pages 356-365.

[LAND71] B.S. Landman and R.L. Rosso, "On a Pin Versus Block Relationship for Partitions of Logic Graphs", *IEEE Transactions on Computers*, C-20, 1971, pages 1469-1479.

[LAVI02] M. Lavin and L. Liebmann, "CAD Computation for Manufacturability: Can We Save VLSI Technology from Itself?", *International Conference on Computer Aided Design*, Nov 2002, pages 424-431.

[LEVE94] M.D. Levenson, "Extending the Lifetime of Optical Lithography Technologies with Wavefront Engineering", *Journal of Applied Physics*, vol.3, 1994, pages 6765.

[LGSY91] http://www.cbl.ncsu.edu/pub/Benchmark_dirs/LGSynth91/

[LIEB03] L. W. Liebmann, "Layout impact of resolution enhancement techniques: impediment or opportunity?", *International Symposium on Physical Design*, Apr 2003, pages.110-117.

[MCNC] MCNC benchmark, http://www.cbl.ncsu.edu/benchmarks/

[MCNC92] MCNC92 benchmark, http://www.cbl.ncsu.edu/CBL_Docs/lys92.html

[MO00] F. Mo, A. Tabbara and R.K. Brayton, "A Force-Directed Macro-Cell Placer", *International Conference on Computer Aided Design*, Nov 2000, pages 177-180.

[MO02a] F. Mo and R.K. Brayton, "Regular Fabrics in Deep Sub-Micron Integrated-Circuit Design", *International Workshop on Logic and Synthesis*, May 2002, pages 7-12.

[MO02b] F. Mo and R.K. Brayton, "River PLA: A Regular Circuit Structure", *Design Automation Conference*, Jun 2002, pages 201-206.

[MO03a] F. Mo and R.K. Brayton, "Fishbone: A Block-Level Placement and Routing Scheme", *International Symposium on Physical Design*, Apr 2003, pages 204-209.

[MO03b] F. Mo and R.K. Brayton, "PLA-Based Regular Structures and Their Synthesis", *IEEE Transactions on Computer-Aided Design of Integrated Circuits and Systems*, Jun 2003, pages 723-729.

[MO03c] F. Mo and R. K. Brayton, "Checkerboard: A Regular Structure and Its Synthesis", *International Workshop on Logic and Synthesis*, May 2003.

[MULL86] R.S. Muller and T.I. Kamins, *Device Electronics for Integrated Circuits*, Wiley, New York, 1986.

[MURA96] H. Murata, K. Fujiyoshi, S. Nakatake and Y. Kajitani, "VLSI Module Placement Based on Rectangle-Packing by the Sequence-Pair", *IEEE Transactions on Computer-Aided Design of Integrated Circuits and Systems*, vol. 15, no. 12, Dec 1996, pages 1518-1524.

[ONG85] D.G. Ong, *Modern MOS Technology, Process, Devices and Design*, McGraw-Hill Inc., New York, 1984.

[OU00] S-L. Ou and M. Pedram, "Timing-Driven Placement Based on Partitioning with Dynamic Cut-net Control", *Design Automation Conference*, 2000, pages 472-476.

[PAPA90] C. Papachristou and A. Pandya, "A design scheme for PLA-based control tables with reduced area and time-delay cost", *IEEE Transactions on Computer Aided-Design*, vol. 9, no. 5, May 1990, pages 453-472.

[PATE03] C. Patel, A. Cozzie, H. Schmit and L. Pileggi, "An Architectural Exploration of Via Patterned Gate Arrays", *International Symposium of Physical Design*, Apr 2003, pages 184-189.

[PLUM00] J.D. Plummer, M.D. Deal and P.B. Griffin, *Silicon VLSI Technology*, chapter 5, Prentice Hall, 2000.

[QUIN79] N.R. Quinn, JR and M.A. Breuer, "A Force Directed Component Placement Procedure for Printed Circuit Boards", *IEEE Transactions on Circuit and System*, vol. 26, no. 6, 1979, pages 377-388.

[RUDE87] R. Rudell and A. Sangiovanni-Vincentelli, "Multiple-valued minimization for PLA optimization", *IEEE Transactions on Computer Aided-Design*, vol. 6, Sep 1987, pages 727-750.

[SAIT99] S.M. Sait and H. Youssef, *VLSI Physical Design Automation, Theory and Practice*, edited by World Scientific, 1999.

[SECH88] C. Sechen, "Chip-planning, Placement and Global Routing of Macro/custom Cell Integrated Circuits using Simulated Annealing", Design Automation Conference, 1988, pages 73-80.

[SENT92] E. Sentovich, K. Singh, L. Lavagno, C. Moon, R. Murgai, A. Saldanha, H. Savoj, P. Stephan, R. Brayton and A. Sangiovanni-Vincentelli, *SIS: A system for sequential circuit synthesis*, Tech. Rep., UCB/ERL M92/41, Electronics Research Lab, University of California, Berkeley, May 1992.

[SHER93] N.A.Sherwani, *Algorithms for Physical Design Automation*, Kluwer Academic, 1993.

[SHER95] N. Sherwani, S. Bhingarde and a. panyam, *Routing in the Third Dimension from VLSI to MCMs*, IEEE Press. 1995.

[SHIR96] H. Shirota, S. Shibatani, and M. Terai, "A New Rip-up and Reroute Algorithm for Very Large Scale Gate Arrays", *Custom Integrated-Circuits Conference*, 1996, pages 9.3.1-9.3.4.

[SING01] A. Singh, G. Parthasarthy and M. Marek-Sadowska, "Interconnect Resource-Aware Placement for Hierarchical FPGAs", *International Conference on Computer-aided Design*, Nov 2001, pages 132-136.

[SMIT97] S. Smith, "Application Specific Integrated Circuits", Addison-Wesley, 1st edition, 1997.

[STRE80] B.G. Streetman, *Solid-State Electronic Devices*, Prentice-Hall, Englewood Cliffs, New Jersey, 1980.

[SYNO] http://www.synopsys.com/products/unified_synthesis/unified_synthesis.html

[SZE85] S.M. Sze, *Semiconductor Devices: Physics and Technology*, Wiley, New York, 1985.

[TABB98] A. Tabbara, *Retiming for DSM with Area-Delay Trade-Offs and Delay Constraints*, M.S. thesis, Department of Electrical Engineering and Computer Sciences, University of California Berkeley, 1998.

[TZEN88] P.S. Tzeng and C.H. Sequin, "Codar: A Congestion-Directed General Area Router", *International Conference on Computer Aided Design*, 1988, pages 30-33.

[VANG90] L.P.P.P. van Ginneken, "Buffer Placement in Distributed RC-Tree Networks for Minimal Elmore Delay", *International Symposium on Circuits and Systems*, 1990, pages 865-868.

[VYGE97] J. Vygen, "Algorithms for Large-Scale Flat Placement", *Design Automation Conference*, 1997, pages 746-751.

[WANG00] M. Wang, X. Yang and M. Sarrafzadeh, "Dragon 2000: Standard-cell Placement Tool for Large Industry Circuits", *International Conference on Computer Aided Design*, 2000, pages 260-263.

[WANG01] J-S. Wang, C-R. Chang and C. Yeh, "Analysis and Design of High-Speed and Low-Power CMOS PLAs", *IEEE Journal of Solid-State Circuits*, vol. 36, no. 8, Aug 2001, pages 1250-1262.

[WEST94] N.H.E. Weste and K. Eshraghian, *Principle of CMOS VLSI Design*, Addison-Wesley Pub. Co., 1994.

[XILI99] Xilinx Datasheet, XC4000E and XC4000X series field programmable gate arrays, version 1.6, May 1999.

[YANG88] E.S. Yang, *Microelectronic Devices*, McGraw-Hill, New York, 1988.

[YANG02] X. Yang, B-K. Choi and M. Sarrafzadeh, "Routability Driven White Space Allocation for Fixed-Die Standard-Cell Placement", *International Symposium on Physical Design*, 2002, pages 42-50.

[YILD01] M.C. Yildiz and P.H. Madden, "Improved Cut Sequences for Partitioning Based Placement", *Design Automation Conference*, 2001, pages 776-779.

[YOSH01] H. Yoshimura, Y. Asahi and F. Matsuoka, "Scaling Scenario of Multi-level Interconnects for Future CMOS LSI", *Digest of Technical Papers, Symposium on VLSI Technology*, 2001, pages 143-144.

[YOUN03] E.F.Y. Young, C.C.N. Chu and Z.C. Shen, "Twin Binary Sequences: A Nonredundant Representation for General Nonsliceing Floorplan", *IEEE Transactions on Computer-Aided Design*, vol. 22, no. 4, Apr 2003, pages 457-469.

[ZHAN00] S. Zhang and W. Dai, "Linear Time Left Edge Algorithm", *International Conference on Chip Design Automation*, Beijing, China, Aug 2000.

INDEX

R

S

T

W

MIX
Papier aus verantwortungsvollen Quellen
Paper from responsible sources
FSC® C105338

If you have any concerns about our products,
you can contact us on
ProductSafety@springernature.com

In case Publisher is established outside the EU,
the EU authorized representative is:
Springer Nature Customer Service Center GmbH
Europaplatz 3, 69115 Heidelberg, Germany

Printed by Libri Plureos GmbH
in Hamburg, Germany